COFFIN'S
INTEREST TABLES

ONE DOLLAR TO TEN THOUSAND DOLLARS

Including a Master Table Showing Interest at
Various Rates From One-eighth Per Cent to
Ten Per Cent

•

TIMETABLE · COMPOUND INTEREST TABLE
INTEREST LAWS · DIGEST OF BUSINESS
LAWS · BUSINESS FORMS · POSTAL RATES

•

REVISED EDITION

HOLT, RINEHART AND WINSTON
New York • Chicago • San Francisco

ISBN: 0-03-037285-2
PRINTED IN THE UNITED STATES OF AMERICA

FOREWORD

This new edition of COFFIN'S INTEREST TABLES has been revised to make it more serviceable to the large number of clients who have been constant users of the book since its first publication.

The book has been brought up-to-date by the addition of new material. Fractional interest rates beginning with ⅛ per cent on one dollar have been included in order to meet the demand for computation at the prevailing lower percentages.

Consequently, the scope of its field of use has been extended to include banks, finance and bond houses, mortgage and note brokers, public accountants, auditors, income tax experts, and many others in the field of business finance whose accountants require the aid of a reliable publication such as Coffin's—for years an authority in this field.

HOW TO USE THIS BOOK

This revised edition of COFFIN'S INTEREST TABLES provides two methods for ascertaining the interest on any sum. Your choice of method should be based upon the rate involved.

If the rate is $\frac{1}{2}$, $3\frac{1}{2}$, $4\frac{1}{2}$ *or any whole number per cent*, time is saved by using the individual tables arranged by amounts of principal from one dollar to ten thousand dollars.

If the rate is a fractional per cent, use the Master Table (pages 4 to 9), which shows the interest on one dollar for twenty-four various rates and, by means of which, the interest on any sum may be calculated. For accuracy the figures in the table are extended to five decimal places.

EXAMPLE I

Required, the interest on $485 for one year, four months, twenty-eight days at $3\frac{1}{8}$ per cent.

Turn to the Master Table, page 4. The computation of the interest on

$1 at 3 per cent for 1 year is	.03000
for 4 months	.01000
for 28 days	.00233
Interest on $1 at ⅛ per cent for 1 year	.00125
for 4 months	.00042
for 28 days	.00010

The sum of these items is the interest on $1 for the stated period .	.04410
Multiplying by the stated amount	485

22050
35280
17640

We have the interest required: $21.38850

2

FOREWORD

EXAMPLE II

Required, the interest on $150,000 for 64 days at 2¼ per cent.

On page 7 of the Master Table, the interest on $1 at 2¼ per cent
for 64 days is given as 00400
Multiplying the stated sum $150000
by the interest on $1 00400

We have the interest required: $600.00000

EXAMPLE III

Required, the interest on $95 for 3 years, 6 months, 15 days at 4½ per cent.

Turning to page 104 we find the interest on $95 at 4½ per cent
for 3 years is $12.83
for 6 months 2.14
for 15 days 18

for the required time $15.15

EXAMPLE IV

Required, the interest on $275 for 5 years, 6 months at 5 per cent.

Turning to page 110, we find the interest on $200 at 5 per cent
for 5 years is $50.00
for 6 months 5.00

On page 84 we find the interest on $75
for 5 years is $18.75
for 6 months 1.88

The required interest is the sum of the interest on $200 and $75 for the
required time $75.63

COMPOUND INTEREST

The Compound Interest Table, page 130, shows the amount of one dollar compounded annually from 1 to 20 years at rates of interest from ½ per cent to 8 per cent. To find what any sum will amount to, by referring to the Table, determine the compound amount on $1, and multiply this by the sum.

EXAMPLE I

Required, the amount of $5,156 at compound interest for 9 years at 2½ per cent.

From the Compound Interest Table, page 130, we find that the
amount of $1 for 9 years at 2½ per cent is $1.248863
Multiplying by the given amount   5156

  7493178
  6244315
  1248863
  6244315

The required amount is $6,439.137628

FOREWORD

EXAMPLE II

Required, the amount of $2,750 at compound interest for 12 years, at 3½ per cent.
From the Compound Interest Table, page 130, we find that the
amount of $1 for 12 years at 3½ per cent is $1.511069
Multiplying by the given amount   2750

                                                            75553450
                                                            10577483
                                                             3022138

We have the required amount $4,155.439750

THE TIMETABLE

The table, pages 128–129, provides a ready means of finding the number of days between any two given dates. A year from any date is the corresponding day of the following year. A month from any date is the corresponding day of the following month; but it cannot extend beyond the last day of the following month. A month from December 28 is January 28, but a month from January 28, January 29, January 30, or January 31 is February 28 in a common year (February 29 in a leap year). Two months from December 30 is the last day of the following February. However, reckoned by days, thirty days from January 31 is March 2 in a common year, March 1 in a leap year.

EXAMPLE I

Find the number of days from March 20 to July 27. In the table, page 128, under March, we find beneath March 20 the number 79, and in the July column, the number 208 beneath July 27. From 208 subtract 79; the difference is the number of days required, 129.

EXAMPLE II

Find the number of days between August 17 and April 30. In the table, page 128, the number, 229, is found under August 17, and, page 129 (the second year), the number, 485, is underneath April 30. The difference between these two numbers, 256, is the answer. Should February intervene in a leap year, one day is added to the answer.

To reduce the number of days to months, divide by 30

            Thus: 30)256(8
                       240
                       —
                    16   8 months, 16 days

MASTER TABLE
One Dollar at Various Per Cents

Years.	⅛ per ct.	¼ per ct.	⅜ per ct.	½ per ct.	⅝ per ct.	¾ per ct.	⅞ per ct.	1 per ct.	1¼ per ct.	1½ per ct.	1¾ per ct.	2 per ct.
1	.00125	.00250	.00375	.00500	.00625	.00750	.00875	.01000	.01250	.01500	.01750	.02000
2	.00250	.00500	.00750	.01000	.01250	.01500	.01750	.02000	.02500	.03000	.03500	.04000
3	.00375	.00750	.01125	.01500	.01875	.02250	.02625	.03000	.03750	.04500	.05250	.06000
4	.00500	.01000	.01500	.02000	.02500	.03000	.03500	.04000	.05000	.06000	.07000	.08000
5	.00625	.01250	.01875	.02500	.03125	.03750	.04375	.05000	.06250	.07500	.08750	.10000
Months.												
1	.00010	.00021	.00031	.00042	.00052	.00063	.00073	.00083	.00104	.00125	.00146	.00167
2	.00021	.00042	.00063	.00083	.00104	.00125	.00146	.00167	.00208	.00250	.00292	.00333
3	.00031	.00063	.00094	.00125	.00156	.00188	.00219	.00250	.00313	.00375	.00438	.00500
4	.00042	.00083	.00125	.00167	.00208	.00250	.00292	.00333	.00417	.00500	.00583	.00667
5	.00052	.00104	.00156	.00208	.00260	.00313	.00365	.00417	.00521	.00625	.00729	.00833
6	.00063	.00125	.00188	.00250	.00313	.00375	.00438	.00500	.00625	.00750	.00875	.01000
7	.00073	.00146	.00219	.00292	.00365	.00438	.00510	.00583	.00729	.00875	.01021	.01167
8	.00083	.00167	.00250	.00333	.00417	.00500	.00583	.00667	.00833	.01000	.01167	.01333
9	.00094	.00188	.00281	.00375	.00469	.00563	.00656	.00750	.00938	.01125	.01313	.01500
10	.00104	.00208	.00313	.00417	.00521	.00625	.00729	.00833	.01042	.01250	.01458	.01667
11	.00115	.00229	.00344	.00458	.00573	.00688	.00802	.00917	.01146	.01375	.01604	.01833
Days.												
1	.00000	.00001	.00001	.00001	.00002	.00002	.00002	.00003	.00004	.00004	.00005	.00006
2	.00001	.00001	.00002	.00003	.00004	.00004	.00005	.00006	.00007	.00008	.00010	.00011
3	.00001	.00002	.00003	.00004	.00005	.00006	.00007	.00008	.00010	.00013	.00015	.00017
4	.00001	.00003	.00004	.00006	.00007	.00008	.00010	.00011	.00014	.00017	.00019	.00022
5	.00002	.00004	.00005	.00007	.00009	.00010	.00012	.00014	.00017	.00021	.00024	.00028
6	.00002	.00004	.00006	.00008	.00010	.00013	.00015	.00017	.00021	.00025	.00029	.00033
7	.00002	.00005	.00007	.00010	.00012	.00015	.00017	.00019	.00024	.00029	.00034	.00039
8	.00003	.00006	.00008	.00011	.00014	.00017	.00019	.00022	.00028	.00033	.00039	.00044
9	.00003	.00006	.00009	.00013	.00016	.00019	.00022	.00025	.00031	.00038	.00044	.00050
10	.00004	.00007	.00010	.00014	.00017	.00021	.00024	.00028	.00035	.00042	.00049	.00056
11	.00004	.00008	.00012	.00015	.00019	.00023	.00027	.00031	.00038	.00046	.00054	.00061
12	.00004	.00008	.00013	.00017	.00021	.00025	.00029	.00033	.00042	.00050	.00058	.00067
13	.00005	.00009	.00014	.00018	.00023	.00027	.00032	.00036	.00045	.00054	.00063	.00072
14	.00005	.00010	.00015	.00019	.00024	.00029	.00034	.00039	.00049	.00058	.00068	.00078
15	.00005	.00010	.00016	.00021	.00026	.00031	.00037	.00042	.00052	.00063	.00073	.00083
16	.00006	.00011	.00017	.00022	.00028	.00033	.00039	.00044	.00056	.00067	.00078	.00089
17	.00006	.00012	.00018	.00024	.00030	.00035	.00041	.00047	.00059	.00071	.00083	.00094
18	.00006	.00013	.00019	.00025	.00031	.00038	.00044	.00050	.00063	.00075	.00088	.00100
19	.00007	.00013	.00020	.00026	.00033	.00040	.00046	.00053	.00066	.00079	.00092	.00106
20	.00007	.00014	.00021	.00028	.00035	.00042	.00049	.00056	.00069	.00083	.00097	.00111
21	.00007	.00015	.00022	.00029	.00037	.00044	.00051	.00058	.00073	.00088	.00102	.00117
22	.00008	.00015	.00023	.00031	.00038	.00046	.00054	.00061	.00076	.00092	.00107	.00122
23	.00008	.00016	.00024	.00032	.00040	.00048	.00056	.00064	.00080	.00096	.00112	.00128
24	.00008	.00017	.00025	.00033	.00042	.00050	.00058	.00067	.00083	.00100	.00117	.00133
25	.00009	.00017	.00026	.00035	.00043	.00052	.00061	.00069	.00087	.00104	.00122	.00139
26	.00009	.00018	.00027	.00036	.00045	.00054	.00063	.00072	.00090	.00108	.00126	.00144
27	.00009	.00019	.00028	.00038	.00047	.00056	.00066	.00075	.00094	.00113	.00131	.00150
28	.00010	.00019	.00029	.00039	.00049	.00058	.00068	.00078	.00097	.00117	.00136	.00156
29	.00010	.00020	.00030	.00040	.00050	.00060	.00071	.00081	.00101	.00121	.00141	.00161
30	.00010	.00021	.00031	.00042	.00052	.00063	.00073	.00083	.00104	.00125	.00146	.00167

MASTER TABLE
One Dollar at Various Per Cents

	2¼ per ct.	2½ per ct.	2¾ per ct.	3 per ct.	3½ per ct.	4 per ct.	4½ per ct.	5 per ct.	6 per ct.	7 per ct.	8 per ct.	10 per ct.
Years.												
1	.02250	.02500	.02750	.03000	.03500	.04000	.04500	.05000	.06000	.07000	.08000	.10000
2	.04500	.05000	.05500	.06000	.07000	.08000	.09000	.10000	.12000	.14000	.16000	.20000
3	.06750	.07500	.08250	.09000	.10500	.12000	.13500	.15000	.18000	.21000	.24000	.30000
4	.09000	.10000	.11000	.12000	.14000	.16000	.18000	.20000	.24000	.28000	.32000	.40000
5	.11250	.12500	.13750	.15000	.17500	.20000	.22500	.25000	.30000	.35000	.40000	.50000
Months.												
1	.00188	.00208	.00229	.00250	.00292	.00333	.00375	.00417	.00500	.00583	.00667	.00833
2	.00375	.00417	.00458	.00500	.00583	.00667	.00750	.00833	.01000	.01167	.01333	.01667
3	.00563	.00625	.00688	.00750	.00875	.01000	.01125	.01250	.01500	.01750	.02000	.02500
4	.00750	.00833	.00917	.01000	.01167	.01333	.01500	.01667	.02000	.02333	.02667	.03333
5	.00938	.01042	.01146	.01250	.01458	.01667	.01875	.02083	.02500	.02917	.03333	.04167
6	.01125	.01250	.01375	.01500	.01750	.02000	.02250	.02500	.03000	.03500	.04000	.05000
7	.01313	.01458	.01604	.01750	.02042	.02333	.02625	.02917	.03500	.04083	.04667	.05833
8	.01500	.01667	.01833	.02000	.02333	.02667	.03000	.03333	.04000	.04667	.05333	.06667
9	.01688	.01875	.02063	.02250	.02625	.03000	.03375	.03750	.04500	.05250	.06000	.07500
10	.01875	.02083	.02292	.02500	.02917	.03333	.03750	.04167	.05000	.05833	.06667	.08333
11	.02063	.02292	.02521	.02750	.03208	.03667	.04125	.04583	.05500	.06417	.07333	.09167
Days.												
1	.00006	.00007	.00008	.00008	.00010	.00011	.00013	.00014	.00017	.00019	.00022	.00028
2	.00013	.00014	.00015	.00017	.00019	.00022	.00025	.00028	.00033	.00039	.00044	.00056
3	.00019	.00021	.00023	.00025	.00029	.00033	.00038	.00042	.00050	.00058	.00067	.00083
4	.00025	.00028	.00031	.00033	.00039	.00044	.00050	.00056	.00067	.00078	.00089	.00111
5	.00031	.00035	.00038	.00042	.00049	.00056	.00063	.00069	.00083	.00097	.00111	.00139
6	.00038	.00042	.00046	.00050	.00058	.00067	.00075	.00083	.00100	.00117	.00133	.00167
7	.00044	.00049	.00054	.00058	.00068	.00078	.00088	.00097	.00117	.00136	.00156	.00194
8	.00050	.00056	.00061	.00067	.00078	.00089	.00100	.00111	.00133	.00156	.00178	.00222
9	.00056	.00063	.00069	.00075	.00088	.00100	.00113	.00125	.00150	.00175	.00200	.00250
10	.00063	.00069	.00076	.00083	.00097	.00111	.00125	.00139	.00167	.00194	.00222	.00278
11	.00069	.00076	.00084	.00092	.00107	.00122	.00138	.00153	.00183	.00214	.00244	.00306
12	.00075	.00083	.00092	.00100	.00117	.00133	.00150	.00167	.00200	.00233	.00267	.00333
13	.00081	.00090	.00099	.00108	.00126	.00144	.00163	.00181	.00217	.00253	.00289	.00361
14	.00088	.00097	.00107	.00117	.00136	.00156	.00175	.00194	.00233	.00272	.00311	.00389
15	.00094	.00104	.00115	.00125	.00146	.00167	.00188	.00208	.00250	.00292	.00333	.00417
16	.00100	.00111	.00122	.00133	.00156	.00178	.00200	.00222	.00267	.00311	.00356	.00444
17	.00106	.00118	.00130	.00142	.00165	.00189	.00213	.00236	.00283	.00331	.00378	.00472
18	.00113	.00125	.00138	.00150	.00175	.00200	.00225	.00250	.00300	.00350	.00400	.00500
19	.00119	.00132	.00145	.00158	.00185	.00211	.00238	.00264	.00317	.00369	.00422	.00528
20	.00125	.00139	.00153	.00167	.00194	.00222	.00250	.00278	.00333	.00389	.00444	.00556
21	.00131	.00146	.00160	.00175	.00204	.00233	.00263	.00292	.00350	.00408	.00467	.00583
22	.00138	.00153	.00168	.00183	.00214	.00244	.00275	.00306	.00367	.00428	.00489	.00611
23	.00144	.00160	.00176	.00192	.00224	.00256	.00288	.00319	.00383	.00447	.00511	.00639
24	.00150	.00167	.00183	.00200	.00233	.00267	.00300	.00333	.00400	.00467	.00533	.00667
25	.00156	.00174	.00191	.00208	.00243	.00278	.00313	.00347	.00417	.00486	.00556	.00694
26	.00163	.00181	.00199	.00217	.00253	.00289	.00325	.00361	.00433	.00506	.00578	.00722
27	.00169	.00188	.00206	.00225	.00263	.00300	.00338	.00375	.00450	.00525	.00600	.00750
28	.00175	.00194	.00214	.00233	.00272	.00311	.00350	.00389	.00467	.00544	.00622	.00778
29	.00181	.00201	.00222	.00242	.00282	.00322	.00363	.00403	.00483	.00564	.00644	.00806
30	.00188	.00208	.00229	.00250	.00292	.00333	.00375	.00417	.00500	.00583	.00667	.00833

One Dollar at Various Per Cents

Days.	⅛ per ct.	¼ per ct.	⅜ per ct.	½ per ct.	⅝ per ct.	¾ per ct.	⅞ per ct.	1 per ct.	1¼ per ct.	1½ per ct.	1¾ per ct.	2 per ct.
31	.00011	.00022	.00032	.00043	.00054	.00065	.00075	.00086	.00108	.00129	.00151	.00172
32	.00011	.00022	.00033	.00044	.00056	.00067	.00078	.00089	.00111	.00133	.00156	.00178
33	.00012	.00023	.00034	.00046	.00057	.00069	.00080	.00092	.00115	.00138	.00160	.00183
34	.00012	.00024	.00035	.00047	.00059	.00071	.00083	.00094	.00118	.00142	.00165	.00189
35	.00012	.00024	.00037	.00049	.00061	.00073	.00085	.00097	.00122	.00146	.00170	.00194
36	.00013	.00025	.00038	.00050	.00063	.00075	.00088	.00100	.00125	.00150	.00175	.00200
37	.00013	.00026	.00039	.00051	.00064	.00077	.00090	.00103	.00129	.00154	.00180	.00206
38	.00013	.00026	.00040	.00053	.00066	.00079	.00092	.00106	.00132	.00158	.00185	.00211
39	.00014	.00027	.00041	.00054	.00068	.00081	.00095	.00108	.00135	.00163	.00190	.00217
40	.00014	.00028	.00042	.00056	.00069	.00083	.00097	.00111	.00139	.00167	.00194	.00222
41	.00014	.00029	.00043	.00057	.00071	.00085	.00100	.00114	.00142	.00171	.00199	.00228
42	.00015	.00029	.00044	.00058	.90073	.00088	.00102	.00117	.00146	.00175	.00204	.00233
43	.00015	.00030	.00045	.00060	.00075	.00090	.00105	.00119	.00149	.00179	.00209	.00239
44	.00015	.00031	.00046	.00061	.00076	.00092	.00107	.00122	.00153	.00183	.00214	.00244
45	.00016	.00031	.00047	.00063	.00078	.00094	.00109	.00125	.00156	.00188	.00219	.00250
46	.00016	.00032	.00048	.00064	.00080	.00096	.00112	.00128	.00160	.00192	.00224	.00256
47	.00016	.00033	.00049	.00065	.00082	.00098	.00114	.00131	.00163	.00196	.00229	.00261
48	.00017	.00033	.00050	.00067	.00083	.00100	.00117	.00133	.00167	.00200	.00233	.00267
49	.00017	.00034	.00051	.00068	.00085	.00102	.00119	.00136	.00170	.00204	.00238	.00272
50	.00017	.00035	.00052	.00069	.00087	.00104	.00122	.00139	.00174	.00208	.00243	.00278
51	.00018	.00035	.00053	.00071	.00089	.00106	.00124	.00142	.00177	.00213	.00248	.00283
52	.00018	.00036	.00054	.00072	.00090	.00108	.00126	.00144	.00181	.00217	.00253	.00289
53	.00018	.00037	.00055	.00074	.00092	.00110	.00129	.00147	.00184	.00221	.00258	.00294
54	.00019	.00038	.00056	.00075	.00094	.00113	.00131	.00150	.00188	.00225	.00263	.00300
55	.00019	.00038	.00057	.00076	.00096	.00115	.00134	.00153	.00191	.00229	.00267	.00306
56	.00019	.00039	.00058	.00078	.00097	.00117	.00136	.00156	.00194	.00233	.00272	.00311
57	.00020	.00040	.00059	.00079	.00099	.00119	.00139	.00158	.00198	.00238	.00277	.00317
58	.00020	.00040	.00060	.00081	.00101	.00121	.00141	.00161	.00201	.00242	.00282	.00322
59	.00021	.00041	.00062	.00082	.00102	.00123	.00143	.00164	.00205	.00246	.00287	.00328
60	.00021	.00042	.00063	.00083	.00104	.00125	.00146	.00167	.00208	.00250	.00292	.00333
61	.00021	.00042	.00064	.00085	.00106	.00127	.00148	.00169	.00212	.00254	.00297	.00339
62	.00022	.00043	.00065	.00086	.00108	.00129	.00151	.00172	.00215	.00258	.00301	.00344
63	.00022	.00044	.00066	.00088	.00109	.00131	.00153	.00175	.00219	.00263	.00306	.00350
64	.00022	.00044	.00067	.00089	.00111	.00133	.00156	.00178	.00222	.00267	.00311	.00356
65	.00023	.00045	.00068	.00090	.00113	.00135	.00158	.00181	.00226	.00271	.00316	.00361
66	.00023	.00046	.00069	.00092	.00115	.00138	.00160	.00183	.00229	.00275	.00321	.00367
67	.00023	.00047	.00070	.00093	.00116	.00140	.00163	.00186	.00233	.00279	.00326	.00372
68	.00024	.00047	.00071	.00094	.00118	.00142	.00165	.00189	.00236	.00283	.00331	.00378
69	.00024	.00048	.00072	.00096	.00120	.00144	.00168	.00192	.00240	.00288	.00335	.00383
70	.00024	.00049	.00073	.00097	.00122	.00146	.00170	.00194	.00243	.00292	.00340	.00389
71	.00025	.00049	.00074	.00099	.00123	.00148	.00173	.00197	.00247	.00296	.00345	.00394
72	.00025	.00050	.00075	.00100	.00125	.00150	.00175	.00200	.00250	.00300	.00350	.00400
73	.00025	.00051	.00076	.00101	.00127	.00152	.00177	.00203	.00254	.00304	.00355	.00406
74	.00026	.00051	.00077	.00103	.00129	.00154	.00180	.00206	.00257	.00308	.00360	.00411
75	.00026	.00052	.00078	.00104	.00130	.00156	.00182	.00208	.00260	.00313	.00365	.00417
76	.00026	.00053	.00079	.00106	.00132	.00158	.00185	.00211	.00264	.00317	.00369	.00422
77	.00027	.00054	.00080	.00107	.00134	.00160	.00187	.00214	.00267	.00321	.00374	.00428
78	.00027	.00054	.00081	.00108	.00135	.00163	.00190	.00217	.00271	.00325	.00379	.00433

MASTER TABLE
One Dollar at Various Per Cents

Days.	2¼ per ct.	2½ per ct.	2¾ per ct.	3 per ct.	3½ per ct.	4 per ct.	4½ per ct.	5 per ct.	6 per ct.	7 per ct.	8 per ct.	10 per ct.
31	.00194	.00215	.00237	.00258	.00301	.00344	.00388	.00431	.00517	.00603	.00689	.00861
32	.00200	.00222	.00244	.00267	.00311	.00356	.00400	.00444	.00533	.00622	.00711	.00889
33	.00206	.00229	.00252	.00275	.00321	.00367	.00413	.00458	.00550	.00642	.00733	.00917
34	.00213	.00236	.00260	.00283	.00331	.00378	.00425	.00472	.00567	.00661	.00756	.00944
35	.00219	.00243	.00267	.00292	.00340	.00389	.00438	.00486	.00583	.00681	.00778	.00972
36	.00225	.00250	.00275	.00300	.00350	.00400	.00450	.00500	.00600	.00700	.00800	.01000
37	.00231	.00257	.00283	.00308	.00360	.00411	.00463	.00514	.00617	.00719	.00822	.01028
38	.00238	.00264	.00290	.00317	.00369	.00422	.00475	.00528	.00633	.00739	.00844	.01056
39	.00244	.00271	.00298	.00325	.00379	.00433	.00488	.00542	.00650	.00758	.00867	.01083
40	.00250	.00278	.00306	.00333	.00389	.00444	.00500	.00556	.00667	.00778	.00889	.01111
41	.00256	.00285	.00313	.00342	.00399	.00456	.00513	.00569	.00683	.00797	.00911	.01139
42	.00263	.00292	.00321	.00350	.00408	.00467	.00525	.00583	.00700	.00817	.00933	.01167
43	.00269	.00299	.00329	.00358	.00418	.00478	.00538	.00597	.00717	.00836	.00956	.01194
44	.00275	.00306	.00336	.00367	.00428	.00489	.00550	.00611	.00733	.00856	.00978	.01222
45	.00281	.00313	.00344	.00375	.00438	.00500	.00563	.00625	.00750	.00875	.01000	.01250
46	.00288	.00319	.00351	.00383	.00447	.00511	.00575	.00639	.00767	.00894	.01022	.01278
47	.00294	.00326	.00359	.00392	.00457	.00522	.00588	.00653	.00783	.00914	.01044	.01306
48	.00300	.00333	.00367	.00400	.00467	.00533	.00600	.00667	.00800	.00933	.01067	.01333
49	.00306	.00340	.00374	.00408	.00476	.00544	.00613	.00681	.00817	.00953	.01089	.01361
50	.00313	.00347	.00382	.00417	.00486	.00556	.00625	.00694	.00833	.00972	.01111	.01389
51	.00319	.00354	.00390	.00425	.00496	.00567	.00638	.00708	.00850	.00992	.01133	.01417
52	.00325	.00361	.00397	.00433	.00506	.00578	.00650	.00722	.00867	.01011	.01156	.01444
53	.00331	.00368	.00405	.00442	.00515	.00589	.00663	.00736	.00883	.01031	.01178	.01472
54	.00338	.00375	.00413	.00450	.00525	.00600	.00675	.00750	.00900	.01050	.01200	.01500
55	.00344	.00382	.00420	.00458	.00535	.00611	.00688	.00764	.00917	.01069	.01222	.01528
56	.00350	.00389	.00428	.00467	.00544	.00622	.00700	.00778	.00933	.01089	.01244	.01556
57	.00356	.00396	.00435	.00475	.00554	.00633	.00713	.00792	.00950	.01108	.01267	.01583
58	.00363	.00403	.00443	.00483	.00564	.00644	.00725	.00806	.00967	.01128	.01289	.01611
59	.00369	.00410	.00451	.00492	.00574	.00656	.00738	.00819	.00983	.01147	.01311	.01639
60	.00375	.00417	.00458	.00500	.00583	.00667	.00750	.00833	.01000	.01167	.01333	.01667
61	.00381	.00424	.00466	.00508	.00593	.00678	.00763	.00847	.01017	.01186	.01356	.01694
62	.00388	.00431	.00474	.00517	.00603	.00689	.00775	.00861	.01033	.01206	.01378	.01722
63	.00394	.00438	.00481	.00525	.00613	.00700	.00787	.00875	.01050	.01225	.01400	.01750
64	.00400	.00444	.00489	.00533	.00622	.00711	.00800	.00889	.01067	.01244	.01422	.01778
65	.00406	.00451	.00497	.00542	.00632	.00722	.00813	.00903	.01083	.01264	.01444	.01806
66	.00413	.00458	.00504	.00550	.00642	.00733	.00825	.00917	.01100	.01283	.01467	.01833
67	.00419	.00465	.00512	.00558	.00651	.00744	.00838	.00931	.01117	.01303	.01489	.01861
68	.00425	.00472	.00519	.00567	.00661	.00756	.00850	.00944	.01133	.01322	.01511	.01889
69	.00431	.00479	.00527	.00575	.00671	.00767	.00863	.00958	.01150	.01342	.01533	.01917
70	.00438	.00486	.00535	.00583	.00681	.00778	.00875	.00972	.01167	.01361	.01556	.01944
71	.00444	.00493	.00542	.00592	.00690	.00789	.00888	.00986	.01183	.01381	.01578	.01972
72	.00450	.00500	.00550	.00600	.00700	.00800	.00900	.01000	.01200	.01400	.01600	.02000
73	.00456	.00507	.00558	.00608	.00710	.00811	.00913	.01014	.01217	.01419	.01622	.02028
74	.00463	.00514	.00565	.00617	.00719	.00822	.00925	.01028	.01233	.01439	.01644	.02056
75	.00469	.00521	.00573	.00625	.00729	.00833	.00938	.01042	.01250	.01458	.01667	.02083
76	.00475	.00528	.00581	.00633	.00739	.00844	.00950	.01056	.01267	.01478	.01689	.02111
77	.00481	.00535	.00588	.00642	.00749	.00856	.00963	.01069	.01283	.01497	.01711	.02139
78	.00488	.00542	.00596	.00650	.00758	.00867	.00975	.01083	.01300	.01517	.01733	.02167

MASTER TABLE

One Dollar at Various Per Cents

Days.	$\frac{1}{8}$ per ct.	$\frac{1}{4}$ per ct.	$\frac{3}{8}$ per ct.	$\frac{1}{2}$ per ct.	$\frac{5}{8}$ per ct.	$\frac{3}{4}$ per ct.	$\frac{7}{8}$ per ct.	1 per ct.	$1\frac{1}{4}$ per ct.	$1\frac{1}{2}$ per ct.	$1\frac{3}{4}$ per ct.	2 per ct.
79	.00027	.00055	.00082	.00110	.00137	.00165	.00192	.00219	.00274	.00329	.00384	.00439
80	.00028	.00056	.00083	.00111	.00139	.00167	.00194	.00222	.00278	.00333	.00389	.00444
81	.00028	.00056	.00084	.00113	.00141	.00169	.00197	.00225	.00281	.00338	.00394	.00450
82	.00029	.00057	.00085	.00114	.00142	.00171	.00199	.00228	.00285	.00342	.00399	.00456
83	.00029	.00058	.00087	.00115	.00144	.00173	.00202	.00231	.00288	.00346	.00404	.00461
84	.00029	.00058	.00088	.00117	.00146	.00175	.00204	.00233	.00292	.00350	.00408	.00467
85	.00030	.00059	.00089	.00118	.00148	.00177	.00207	.00236	.00295	.00354	.00413	.00472
86	.00030	.00060	.00090	.00119	.00149	.00179	.00209	.00239	.00299	.00358	.00418	.00478
87	.00030	.00060	.00091	.00121	.00151	.00181	.00212	.00242	.00302	.00363	.00423	.00483
88	.00031	.00061	.00092	.00122	.00153	.00183	.00214	.00244	.00306	.00367	.00428	.00489
89	.00031	.00062	.00093	.00124	.00155	.00185	.00216	.00247	.00309	.00371	.00433	.00494
90	.00031	.00063	.00094	.00125	.00156	.00188	.00219	.00250	.00313	.00375	.00438	.00500
91	.00032	.00063	.00095	.00126	.00158	.00190	.00221	.00253	.00316	.00379	.00442	.00506
92	.00032	.00064	.00096	.00128	.00160	.00192	.00224	.00256	.00319	.00383	.00447	.00511
93	.00032	.00065	.00097	.00129	.00162	.00194	.00226	.00258	.00323	.00388	.00452	.00517
94	.00033	.00065	.00098	.00131	.00163	.00196	.00229	.00261	.00326	.00392	.00457	.00522
95	.00033	.00066	.00099	.00132	.00165	.00198	.00231	.00264	.00330	.00396	.00462	.00528
96	.00033	.00067	.00100	.00133	.00167	.00200	.00233	.00267	.00333	.00400	.00467	.00533
97	.00034	.00067	.00101	.00135	.00168	.00202	.00236	.00269	.00337	.00404	.00472	.00539
98	.00034	.00068	.00102	.00136	.00170	.00204	.00238	.00272	.00340	.00408	.00476	.00544
99	.00034	.00069	.00103	.00138	.00172	.00206	.00241	.00275	.00344	.00413	.00481	.00550
100	.00035	.00069	.00104	.00139	.00174	.00208	.00243	.00278	.00347	.00417	.00486	.00556

One Dollar at Various Per Cents

Days.	2¼ per ct.	2½ per ct.	2¾ per ct.	3 per ct.	3½ per ct.	4 per ct.	4½ per ct.	5 per ct.	6 per ct.	7 per ct.	8 per ct.	10 per ct.
79	.00494	.00549	.00604	.00658	.00768	.00878	.00988	.01097	.01317	.01536	.01756	.02194
80	.00500	.00556	.00611	.00667	.00778	.00889	.01000	.01111	.01333	.01556	.01778	.02222
81	.00506	.00563	.00619	.00675	.00788	.00900	.01013	.01125	.01350	.01575	.01800	.02250
82	.00513	.00569	.00626	.00683	.00797	.00911	.01025	.01139	.01367	.01594	.01822	.02278
83	.00519	.00576	.00634	.00692	.00807	.00922	.01038	.01153	.01383	.01614	.01844	.02306
84	.00525	.00583	.00642	.00700	.00817	.00933	.01050	.01167	.01400	.01633	.01867	.02333
85	.00531	.00590	.00649	.00708	.00826	.00944	.01063	.01181	.01417	.01653	.01889	.02361
86	.00538	.00597	.00657	.00717	.00836	.00956	.01075	.01194	.01433	.01672	.01933	.02389
87	.00544	.00604	.00665	.00725	.00846	.00967	.01088	.01208	.01450	.01692	.01933	.02417
88	.00550	.00611	.00672	.00733	.00856	.00978	.01100	.01222	.01467	.01711	.01956	.02444
89	.00556	.00618	.00680	.00742	.00865	.00989	.01112	.01236	.01483	.01731	.01978	.02472
90	.00563	.00625	.00688	.00750	.00875	.01000	.01125	.01250	.01500	.01750	.02000	.02500
91	.00569	.00632	.00695	.00758	.00885	.01011	.01138	.01264	.01517	.01769	.02022	.02528
92	.00575	.00639	.00703	.00767	.00894	.01022	.01150	.01278	.01533	.01789	.02044	.02556
93	.00581	.00646	.00710	.00775	.00904	.01033	.01163	.01292	.01550	.01808	.02067	.02583
94	.00588	.00653	.00718	.00783	.00914	.01044	.01175	.01306	.01567	.01828	.02089	.02611
95	.00594	.00660	.00726	.00792	.00924	.01056	.01188	.01319	.01583	.01847	.02111	.02639
96	.00600	.00667	.00733	.00800	.00933	.01067	.01200	.01333	.01600	.01867	.02133	.02667
97	.00606	.00674	.00741	.00808	.00943	.01078	.01213	.01347	.01617	.01886	.02156	.02694
98	.00613	.00681	.00749	.00817	.00953	.01089	.01225	.01361	.01633	.01906	.02178	.02722
99	.00619	.00688	.00756	.00825	.00963	.01100	.01238	.01375	.01650	.01925	.02200	.02750
100	.00625	.00694	.00764	.00833	.00972	.01111	.01250	.01389	.01667	.01944	.02222	.02778

1 Dollar.

Years.	½ per ct.	1 per ct.	2 per ct.	3 per ct.	3½ per ct.	4 per ct.	4½ per ct.	5 per ct.	6 per ct.	7 per ct.	8 per ct.	10 per ct.
1	01	01	02	03	04	04	05	05	06	07	08	10
2	01	02	04	06	07	08	09	10	12	14	16	20
3	02	03	06	09	11	12	14	15	18	21	24	30
4	02	04	08	12	14	16	18	20	24	28	32	40
5	03	05	10	15	18	20	23	25	30	35	40	50
Months.												
1	0	0	0	0	0	0	0	0	01	01	01	01
2	0	0	0	01	01	01	01	01	01	01	01	02
3	0	0	01	01	01	01	01	01	02	02	02	03
4	0	0	01	01	01	01	02	02	02	02	03	03
5	0	0	01	01	01	02	02	02	03	03	03	04
6	0	01	01	02	02	02	02	03	03	04	04	05
7	0	01	01	02	02	02	03	03	04	04	05	06
8	0	01	01	02	02	03	03	03	04	05	05	07
9	0	01	01	02	03	03	03	04	05	05	06	08
10	0	01	01	03	03	03	04	04	05	06	07	08
11	0	01	01	03	03	04	04	05	06	06	07	09
Days.												
1	0	0	0	0	0	0	0	0	0	0	0	0
2	0	0	0	0	0	0	0	0	0	0	0	0
3	0	0	0	0	0	0	0	0	0	0	0	0
4	0	0	0	0	0	0	0	0	0	0	0	0
5	0	0	0	0	0	0	0	0	0	0	0	0
6	0	0	0	0	0	0	0	0	0	0	0	0
7	0	0	0	0	0	0	0	0	0	0	0	0
8	0	0	0	0	0	0	0	0	0	0	0	0
9	0	0	0	0	0	0	0	0	0	0	0	0
10	0	0	0	0	0	0	0	0	0	0	0	0
11	0	0	0	0	0	0	0	0	0	0	0	0
12	0	0	0	0	0	0	0	0	0	0	0	0
13	0	0	0	0	0	0	0	0	0	0	0	0
14	0	0	0	0	0	0	0	0	0	0	0	0
15	0	0	0	0	0	0	0	0	0	0	0	0
16	0	0	0	0	0	0	0	0	0	0	0	0
17	0	0	0	0	0	0	0	0	0	0	0	0
18	0	0	0	0	0	0	0	0	0	0	0	01
19	0	0	0	0	0	0	0	0	0	0	0	01
20	0	0	0	0	0	0	0	0	0	0	0	01
21	0	0	0	0	0	0	0	0	0	0	0	01
22	0	0	0	0	0	0	0	0	0	0	0	01
23	0	0	0	0	0	0	0	0	0	0	0	01
24	0	0	0	0	0	0	0	0	0	0	01	01
25	0	0	0	0	0	0	0	0	0	0	01	01
26	0	0	0	0	0	0	0	0	0	0	01	01
27	0	0	0	0	0	0	0	0	0	01	01	01
28	0	0	0	0	0	0	0	0	0	01	01	01
29	0	0	0	0	0	0	0	0	0	01	01	01
30	0	0	0	0	0	0	0	0	01	01	01	01
33	0	0	0	0	0	0	0	0	01	01	01	01
62	0	0	0	01	01	01	01	01	01	01	01	02
93	0	0	01	01	01	01	01	01	02	02	02	03

2 Dollars.

	½ per ct.	1 per ct.	2 per ct.	3 per ct.	3½ per ct.	4 per ct.	4½ per ct.	5 per ct.	6 per ct.	7 per ct.	8 per ct.	10 per ct.
Years.												
1	01	02	04	06	07	08	09	10	12	14	16	20
2	02	04	08	12	14	16	18	20	24	28	32	40
3	03	06	12	18	21	24	27	30	36	42	48	60
4	04	08	16	24	28	32	36	40	48	56	64	80
5	05	10	20	30	35	40	45	50	60	70	80	1.00
Months.												
1	0	0	0	01	01	01	01	01	01	01	01	02
2	0	0	01	01	01	01	02	02	02	02	03	03
3	0	01	01	02	02	02	02	03	03	04	04	05
4	0	01	01	02	02	03	03	03	04	05	05	07
5	0	01	02	03	03	03	04	04	05	06	07	08
6	01	01	02	03	04	04	05	05	06	07	08	10
7	01	01	02	04	04	05	05	06	07	08	09	12
8	01	01	03	04	05	05	06	07	08	09	11	13
9	01	02	03	05	05	06	07	08	09	11	12	15
10	01	02	03	0b	06	07	08	08	10	12	13	17
11	01	02	04	06	06	07	08	09	11	13	15	18
Days.												
1	0	0	0	0	0	0	0	0	0	0	0	0
2	0	0	0	0	0	0	0	0	0	0	0	0
3	0	0	0	0	0	0	0	0	0	0	0	0
4	0	0	0	0	0	0	0	0	0	0	0	0
5	0	0	0	0	0	0	0	0	0	0	0	0
6	0	0	0	0	0	0	0	0	0	0	0	0
7	0	0	0	0	0	0	0	0	0	0	0	0
8	0	0	0	0	0	0	0	0	0	0	0	0
9	0	0	0	0	0	0	0	0	0	0	0	01
10	0	0	0	0	0	0	0	0	0	0	0	01
11	0	0	0	0	0	0	0	0	0	0	0	01
12	0	0	0	0	0	0	0	0	0	0	01	01
13	0	0	0	0	0	0	0	0	0	01	01	01
14	0	0	0	0	0	0	0	0	0	01	01	01
15	0	0	0	0	0	0	0	0	01	01	01	01
16	0	0	0	0	0	0	0	0	01	01	01	01
17	0	0	0	0	0	0	0	0	01	01	01	01
18	0	0	0	0	0	0	0	01	01	01	01	01
19	0	0	0	0	0	0	0	01	01	01	01	01
20	0	0	0	0	0	0	0	01	01	01	01	01
21	0	0	0	0	0	0	01	01	01	01	01	01
22	0	0	0	0	0	0	01	01	01	01	01	01
23	0	0	0	0	0	01	01	01	01	01	01	01
24	0	0	0	0	0	01	01	01	01	01	01	01
25	0	0	0	0	0	01	01	01	01	01	01	01
26	0	0	0	0	01	01	01	01	01	01	01	01
27	0	0	0	0	01	01	01	01	01	01	01	02
28	0	0	0	0	01	01	01	01	01	01	01	02
29	0	0	0	0	01	01	01	01	01	01	01	02
30	0	0	0	01	01	01	01	01	01	01	01	02
33	0	0	0	01	01	01	01	01	01	01	01	02
63	0	0	01	01	01	01	02	02	02	02	03	04
93	0	01	01	02	02	02	02	03	03	04	04	05

3 Dollars.

Years.	½ per ct.	1 per ct.	2 per ct.	3 per ct.	3½ per ct.	4 per ct.	4½ per ct.	5 per ct.	6 per ct.	7 per ct.	8 per ct.	10 per ct.
1	02	03	06	09	11	12	14	15	18	21	24	30
2	03	06	12	18	21	24	27	30	36	42	48	60
3	05	09	18	27	32	36	41	45	54	63	72	90
4	06	12	24	36	42	48	54	60	72	84	96	1.20
5	08	15	30	45	53	60	68	75	90	1.05	1.20	1.50
Months.												
1	0	0	01	01	01	01	01	01	02	02	02	03
2	0	01	01	02	02	02	02	03	03	04	04	05
3	0	01	02	02	03	03	03	04	05	05	06	08
4	01	01	02	03	04	04	05	05	06	07	08	10
5	01	01	03	04	04	05	06	06	08	09	10	13
6	01	02	03	05	05	06	07	08	09	11	12	15
7	01	02	04	05	06	07	08	09	11	12	14	18
8	01	02	04	06	07	08	09	10	12	14	16	20
9	01	02	05	07	08	09	10	11	14	16	18	23
10	01	03	05	08	09	10	11	13	15	18	20	25
11	01	03	06	08	10	11	12	14	17	19	22	28
Days.												
1	0	0	0	0	0	0	0	0	0	0	0	0
2	0	0	0	0	0	0	0	0	0	0	0	0
3	0	0	0	0	0	0	0	0	0	0	0	0
4	0	0	0	0	0	0	0	0	0	0	0	0
5	0	0	0	0	0	0	0	0	0	0	0	0
6	0	0	0	0	0	0	0	0	0	0	0	01
7	0	0	0	0	0	0	0	0	0	0	0	01
8	0	0	0	0	0	0	0	0	0	0	01	01
9	0	0	0	0	0	0	0	0	0	01	01	01
10	0	0	0	0	0	0	0	0	01	01	01	01
11	0	0	0	0	0	0	0	0	01	01	01	01
12	0	0	0	0	0	0	0	01	01	01	01	01
13	0	0	0	0	0	0	0	01	01	01	01	01
14	0	0	0	0	0	0	01	01	01	01	01	01
15	0	0	0	0	0	01	01	01	01	01	01	01
16	0	0	0	0	0	01	01	01	01	01	01	01
17	0	0	0	0	0	01	01	01	01	01	01	01
18	0	0	0	0	01	01	01	01	01	01	01	02
19	0	0	0	0	01	01	01	01	01	01	01	02
20	0	0	0	01	01	01	01	01	01	01	01	02
21	0	0	0	01	01	01	01	01	01	01	01	02
22	0	0	0	01	01	01	01	01	01	01	01	02
23	0	0	0	01	01	01	01	01	01	01	02	02
24	0	0	0	01	01	01	01	01	01	01	02	02
25	0	0	0	01	01	01	01	01	01	01	02	02
26	0	0	0	01	01	01	01	01	01	02	02	02
27	0	0	0	01	01	01	01	01	01	02	02	02
28	0	0	0	01	01	01	01	01	01	02	02	02
29	0	0	0	01	01	01	01	01	01	02	02	02
30	0	0	01	01	01	01	01	01	02	02	02	03
33	0	0	01	01	01	01	01	01	02	02	02	03
63	0	01	01	02	02	02	02	03	03	04	04	05
93	0	01	02	02	03	03	03	04	05	05	06	08

4 Dollars.

	½ per ct.	1 per ct.	2 per ct.	3 per ct.	3½ per ct.	4 per ct.	4½ per ct.	5 per ct.	6 per ct.	7 per ct.	8 per ct.	10 per ct.
Years.												
1	02	04	08	12	14	16	18	20	24	28	32	40
2	04	08	16	24	28	32	36	40	48	56	64	80
3	06	12	24	36	42	48	54	60	72	84	96	1.20
4	08	16	32	48	56	64	72	80	96	1.12	1.28	1.60
5	10	20	40	60	70	80	90	1.00	1.20	1.40	1.60	2.00
Months.												
1	0	0	01	01	01	01	02	02	02	02	03	03
2	0	01	01	02	02	03	03	03	04	05	05	07
3	01	01	02	03	03	04	05	05	06	07	08	10
4	01	01	03	04	05	05	06	07	08	09	11	13
5	01	02	03	05	06	07	08	08	10	12	13	17
6	01	02	04	06	07	08	09	10	12	14	16	20
7	01	02	05	07	08	09	11	12	14	16	19	23
8	01	03	05	08	09	11	12	13	16	19	21	27
9	01	03	06	09	10	12	14	15	18	21	24	30
10	01	03	07	10	12	13	15	17	20	23	27	33
11	01	04	07	11	13	15	17	18	22	26	29	37
Days.												
1	0	0	0	0	0	0	0	0	0	0	0	0
2	0	0	0	0	0	0	0	0	0	0	0	0
3	0	0	0	0	0	0	0	0	0	0	0	0
4	0	0	0	0	0	0	0	0	0	0	0	0
5	0	0	0	0	0	0	0	0	0	0	0	01
6	0	0	0	0	0	0	0	0	0	0	01	01
7	0	0	0	0	0	0	0	0	0	01	01	01
8	0	0	0	0	0	0	0	0	01	01	01	01
9	0	0	0	0	0	0	0	01	01	01	01	01
10	0	0	0	0	0	0	01	01	01	01	01	01
11	0	0	0	0	0	0	01	01	01	01	01	01
12	0	0	0	0	0	01	01	01	01	01	01	01
13	0	0	0	0	01	01	01	01	01	01	01	01
14	0	0	0	0	01	01	01	01	01	01	01	02
15	0	0	0	01	01	01	01	01	01	01	01	02
16	0	0	0	01	01	01	01	01	01	01	01	02
17	0	0	0	01	01	01	01	01	01	01	02	02
18	0	0	0	01	01	01	01	01	01	01	02	02
19	0	0	0	01	01	01	01	01	01	01	02	02
20	0	0	0	01	01	01	01	01	01	02	02	02
21	0	0	0	01	01	01	01	01	01	02	02	02
22	0	0	0	01	01	01	01	01	01	02	02	02
23	0	0	01	01	01	01	01	01	02	02	02	03
24	0	0	01	01	01	01	01	01	02	02	02	03
25	0	0	01	01	01	01	01	01	02	02	02	03
26	0	0	01	01	01	01	01	01	02	02	02	03
27	0	0	01	01	01	01	01	02	02	02	02	03
28	0	0	01	01	01	01	01	02	02	02	02	03
29	0	0	01	01	01	01	01	02	02	02	03	03
30	0	0	01	01	01	01	01	02	02	02	03	03
33	0	0	01	01	01	01	02	02	02	03	03	04
63	0	01	01	02	02	03	03	04	04	05	06	07
93	01	01	02	03	04	04	05	05	06	07	08	10

5 Dollars.

Years.	½ per ct.	1 per ct.	2 per ct.	3 per ct.	3½ per ct.	4 per ct.	4½ per ct.	5 per ct.	6 per ct.	7 per ct.	8 per ct.	10 per ct.
1	03	05	10	15	18	20	23	25	30	35	40	50
2	05	10	20	30	35	40	45	50	60	70	80	1.00
3	08	15	30	45	53	60	68	75	90	1.05	1.20	1.50
4	10	20	40	60	70	80	90	1.00	1.20	1.40	1.60	2.00
5	13	25	50	75	88	1.00	1.13	1.25	1.50	1.75	2.00	2.50
Months.												
1	0	0	01	01	01	02	02	02	03	03	03	04
2	0	01	02	03	03	03	04	04	05	06	07	08
3	01	01	03	04	04	05	06	06	08	09	10	13
4	01	02	03	05	06	07	08	08	10	12	13	17
5	01	02	04	06	07	08	09	10	13	15	17	21
6	01	03	05	08	09	10	11	13	15	18	20	25
7	01	03	06	09	10	12	13	15	18	20	23	29
8	02	03	07	10	12	13	15	17	20	23	27	33
9	02	04	08	11	13	15	17	19	23	26	30	38
10	02	04	08	13	15	17	19	21	25	29	33	42
11	02	05	09	14	16	18	21	23	28	32	37	46
Days.												
1	0	0	0	0	0	0	0	0	0	0	0	0
2	0	0	0	0	0	0	0	0	0	0	0	0
3	0	0	0	0	0	0	0	0	0	0	0	0
4	0	0	0	0	0	0	0	0	0	0	0	01
5	0	0	0	0	0	0	0	0	0	0	01	01
6	0	0	0	0	0	0	0	0	01	01	01	01
7	0	0	0	0	0	0	0	0	01	01	01	01
8	0	0	0	0	0	0	01	01	01	01	01	01
9	0	0	0	0	0	01	01	01	01	01	01	01
10	0	0	0	0	0	01	01	01	01	01	01	01
11	0	0	0	0	01	01	01	01	01	01	01	02
12	0	0	0	01	01	01	01	01	01	01	01	02
13	0	0	0	01	01	01	01	01	01	01	01	02
14	0	0	0	01	01	01	01	01	01	01	02	02
15	0	0	0	01	01	01	01	01	01	01	02	02
16	0	0	0	01	01	01	01	01	01	02	02	02
17	0	0	0	01	01	01	01	01	01	02	02	02
18	0	0	01	01	01	01	01	01	02	02	02	03
19	0	0	01	01	01	01	01	01	02	02	02	03
20	0	0	01	01	01	01	01	01	02	02	02	03
21	0	0	01	01	01	01	01	01	02	02	02	03
22	0	0	01	01	01	01	01	02	02	02	02	03
23	0	0	01	01	01	01	01	02	02	02	03	03
24	0	0	01	01	01	01	02	02	02	02	03	03
25	0	0	01	01	01	01	02	02	02	02	03	03
26	0	0	01	01	01	01	02	02	02	03	03	04
27	0	0	01	01	01	02	02	02	02	03	03	04
28	0	0	01	01	01	02	02	02	02	03	03	04
29	0	0	01	01	01	02	02	02	02	03	03	04
30	0	0	01	01	01	02	02	02	03	03	03	04
33	0	0	01	01	02	02	02	02	03	03	04	05
63	0	01	02	03	03	04	04	04	05	06	07	09
93	01	01	03	04	05	05	06	06	08	09	10	13

6 Dollars.

Years.	½ per ct.	1 per ct.	2 per ct.	3 per ct.	3½ per ct.	4 per ct.	4½ per ct.	5 per ct.	6 per ct.	7 per ct.	8 per ct.	10 per ct.
1	03	06	12	18	21	24	27	30	36	42	48	60
2	06	12	24	36	42	48	54	60	72	84	96	1.20
3	09	18	36	54	63	72	81	90	1.08	1.26	1.44	1.80
4	12	24	48	72	84	96	1.08	1.20	1.44	1.68	1.92	2.40
5	15	30	60	90	1.05	1.20	1.35	1.50	1.80	2.10	2.40	3.00
Months.												
1	0	01	01	02	02	02	02	03	03	04	04	05
2	01	01	02	03	04	04	05	05	06	07	08	10
3	01	02	03	05	05	06	07	08	09	11	12	15
4	01	02	04	06	07	08	09	10	12	14	16	20
5	01	03	05	08	09	10	11	13	15	18	20	25
6	02	03	06	09	11	12	14	15	18	21	24	30
7	02	04	07	11	12	14	16	18	21	25	28	35
8	02	04	08	12	14	16	18	20	24	28	32	40
9	02	05	09	14	16	18	20	23	27	32	36	45
10	03	05	10	15	18	20	23	25	30	35	40	50
11	03	06	11	17	19	22	25	28	33	39	44	55
Days.												
1	0	0	0	0	0	0	0	0	0	0	0	0
2	0	0	0	0	0	0	0	0	0	0	0	0
3	0	0	0	0	0	0	0	0	0	0	0	01
4	0	0	0	0	0	0	0	0	0	0	01	01
5	0	0	0	0	0	0	0	01	01	01	01	01
6	0	0	0	0	0	0	01	01	01	01	01	01
7	0	0	0	0	0	0	01	01	01	01	01	01
8	0	0	0	0	0	01	01	01	01	01	01	01
9	0	0	0	0	01	01	01	01	01	01	01	02
10	0	0	0	01	01	01	01	01	01	01	01	02
11	0	0	0	01	01	01	01	01	01	01	01	02
12	0	0	0	01	01	01	01	01	01	01	02	02
13	0	0	0	01	01	01	01	01	01	02	02	02
14	0	0	0	01	01	01	01	01	01	02	02	02
15	0	0	01	01	01	01	01	01	02	02	02	03
16	0	0	01	01	01	01	01	01	02	02	02	03
17	0	0	01	01	01	01	01	01	02	02	02	03
18	0	0	01	01	01	01	01	02	02	02	02	03
19	0	0	01	01	01	01	01	02	02	02	03	03
20	0	0	01	01	01	01	02	02	02	02	03	03
21	0	0	01	01	01	01	02	02	02	02	03	04
22	0	0	01	01	01	01	02	02	02	03	03	04
23	0	0	01	01	01	02	02	02	02	03	03	04
24	0	0	01	01	01	02	02	02	03	03	03	04
25	0	0	01	01	01	02	02	02	03	03	03	04
26	0	0	01	01	02	02	02	02	03	03	03	04
27	0	0	01	01	02	02	02	02	03	03	04	05
28	0	0	01	01	02	02	02	02	03	03	04	05
29	0	0	01	01	02	02	02	02	03	03	04	05
30	0	01	01	02	02	02	02	03	03	04	04	05
33	0	01	01	02	02	02	02	03	03	04	04	06
63	01	01	02	03	04	04	05	05	06	07	08	11
93	01	02	03	05	05	06	07	08	09	11	12	16

7 Dollars.

Years.	½ per ct.	1 per ct.	2 per ct.	3 per ct.	3½ per ct.	4 per ct.	4½ per ct.	5 per ct.	6 per ct.	7 per ct.	8 per ct.	10 per ct.
1	04	07	14	21	25	28	32	35	42	49	56	70
2	07	14	28	42	49	56	63	70	84	98	1.12	1.40
3	11	21	42	63	74	84	95	1.05	1.26	1.47	1.68	2.10
4	14	28	56	84	98	1.12	1.26	1.40	1.68	1.96	2.24	2.80
5	18	35	70	1.05	1.23	1.40	1.58	1.75	2.10	2.45	2.80	3.50
Months.												
1	0	01	01	02	02	02	03	03	04	04	05	06
2	01	01	02	04	04	05	05	06	07	08	09	12
3	01	02	04	05	06	07	08	09	11	12	14	18
4	01	02	05	07	08	09	11	12	14	16	19	23
5	01	03	06	09	10	12	13	15	18	20	23	29
6	02	04	07	11	12	14	16	18	21	25	28	35
7	02	04	08	12	14	16	18	20	25	29	33	41
8	02	05	09	14	16	19	21	23	28	33	37	47
9	03	05	11	16	18	21	24	26	32	37	42	53
10	03	06	12	18	20	23	26	29	35	41	47	58
11	03	06	13	19	22	26	29	32	39	45	51	64
Days.												
1	0	0	0	0	0	0	0	0	0	0	0	0
2	0	0	0	0	0	0	0	0	0	0	0	0
3	0	0	0	0	0	0	0	0	0	0	0	01
4	0	0	0	0	0	0	0	0	0	01	01	01
5	0	0	0	0	0	0	0	0	01	01	01	01
6	0	0	0	0	0	0	01	01	01	01	01	01
7	0	0	0	0	0	01	01	01	01	01	01	01
8	0	0	0	0	01	01	01	01	01	01	01	02
9	0	0	0	01	01	01	01	01	01	01	01	02
10	0	0	0	01	01	01	01	01	01	01	02	02
11	0	0	0	01	01	01	01	01	01	01	02	02
12	0	0	0	01	01	01	01	01	01	02	02	02
13	0	0	01	01	01	01	01	01	02	02	02	03
14	0	0	01	01	01	01	01	01	02	02	02	03
15	0	0	01	01	01	01	01	01	02	02	02	03
16	0	0	01	01	01	01	01	02	02	02	02	03
17	0	0	01	01	01	01	01	02	02	02	03	03
18	0	0	01	01	01	01	02	02	02	02	03	04
19	0	0	01	01	01	01	02	02	02	03	03	04
20	0	0	01	01	01	02	02	02	02	03	03	04
21	0	0	01	01	01	02	02	02	02	03	03	04
22	0	0	01	01	01	02	02	02	03	03	03	04
23	0	0	01	01	02	02	02	02	03	03	04	04
24	0	0	01	01	02	02	02	02	03	03	04	05
25	0	0	01	01	02	02	02	02	03	03	04	05
26	0	0	01	01	02	02	02	03	03	04	04	05
27	0	01	01	02	02	02	02	03	03	04	04	05
28	0	01	01	02	02	02	02	03	03	04	04	05
29	0	01	01	02	02	02	03	03	03	04	05	06
30	0	01	01	02	02	02	03	03	04	04	05	06
33	0	01	01	02	02	03	03	03	04	04	05	06
63	01	01	02	04	04	05	06	06	07	09	10	12
93	01	02	04	05	06	07	08	09	11	13	14	18

8 Dollars.

Years.	½ per ct.	1 per ct.	2 per ct.	3 per ct.	3½ per ct.	4 per ct.	4½ per ct.	5 per ct.	6 per ct.	7 per ct.	8 per ct.	10 per ct.
1	04	08	16	24	28	32	36	40	48	56	64	80
2	08	16	32	48	56	64	72	80	96	1.12	1.28	1.60
3	12	24	48	72	84	96	1.08	1.20	1.44	1.68	1.92	2.40
4	16	32	64	96	1.12	1.28	1.44	1.60	1.92	2.24	2.56	3.20
5	20	40	80	1.20	1.40	1.60	1.80	2.00	2.40	2.80	3.20	4.00
Months.												
1	0	01	01	02	02	03	03	03	04	05	05	07
2	01	01	03	04	05	05	06	07	08	09	11	13
3	01	02	04	06	07	08	09	10	12	14	16	20
4	01	03	05	08	09	11	12	13	16	19	21	27
5	02	03	07	10	12	13	15	17	20	23	27	33
6	02	04	08	12	14	16	18	20	24	28	32	40
7	02	05	09	14	16	19	21	23	28	33	37	47
8	03	05	11	16	19	21	24	27	32	37	43	53
9	03	06	12	18	21	24	27	30	36	42	48	60
10	03	07	13	20	23	27	30	33	40	47	53	67
11	04	07	15	22	26	29	33	37	44	51	59	73
Days.												
1	0	0	0	0	0	0	0	0	0	0	0	0
2	0	0	0	0	0	0	0	0	0	0	0	0
3	0	0	0	0	0	0	0	0	0	0	01	01
4	0	0	0	0	0	0	0	0	01	01	01	01
5	0	0	0	0	0	0	01	01	01	01	01	01
6	0	0	0	0	0	01	01	01	01	01	01	01
7	0	0	0	0	01	01	01	01	01	01	01	02
8	0	0	0	01	01	01	01	01	01	01	01	02
9	0	0	0	01	01	01	01	01	01	01	02	02
10	0	0	0	01	01	01	01	01	01	02	02	02
11	0	0	0	01	01	01	01	01	01	02	02	02
12	0	0	01	01	01	01	01	01	02	02	02	03
13	0	0	01	01	01	01	01	01	02	02	02	03
14	0	0	01	01	01	01	01	02	02	02	02	03
15	0	0	01	01	01	01	02	02	02	02	03	03
16	0	0	01	01	01	01	02	02	02	02	03	04
17	0	0	01	01	01	02	02	02	02	03	03	04
18	0	0	01	01	01	02	02	02	02	03	03	04
19	0	0	01	01	01	02	02	02	03	03	03	04
20	0	0	01	01	02	02	02	02	03	03	04	04
21	0	0	01	01	02	02	02	02	03	03	04	05
22	0	0	01	01	02	02	02	02	03	03	04	05
23	0	01	01	02	02	02	02	03	03	04	04	05
24	0	01	01	02	02	02	02	03	03	04	04	05
25	0	01	01	02	02	02	03	03	03	04	04	06
26	0	01	01	02	02	02	03	03	03	04	05	06
27	0	01	01	02	02	02	03	03	04	04	05	06
28	0	01	01	02	02	02	03	03	04	04	05	06
29	0	01	01	02	02	03	03	03	04	05	05	06
30	0	01	01	02	02	03	03	03	04	05	05	07
33	0	01	01	02	03	03	03	04	04	05	06	07
63	01	01	03	04	05	06	06	07	08	10	11	14
93	01	02	04	06	07	08	09	10	12	14	17	21

9 Dollars.

Years.	½ per ct.	1 per ct.	2 per ct.	3 per ct.	3½ per ct.	4 per ct.	4½ per ct.	5 per ct.	6 per ct.	7 per ct.	8 per ct.	10 per ct.
1	05	09	18	27	32	36	41	45	54	63	72	90
2	09	18	36	54	63	72	81	90	1.08	1.26	1.44	1.80
3	04	27	54	81	95	1.08	1.22	1.35	1.62	1.89	2.16	2.70
4	18	36	72	1.08	1.26	1.44	1.62	1.80	2.16	2.52	2.88	3.60
5	23	45	90	1.35	1.58	1.80	2.03	2.25	2.70	3.15	3.60	4.50
Months.												
1	0	01	02	02	03	03	03	04	05	05	06	08
2	01	02	03	05	05	06	07	08	09	11	12	15
3	01	02	05	07	08	09	10	11	14	16	18	23
4	02	03	06	09	11	12	14	15	18	21	24	30
5	02	04	08	11	13	15	17	19	23	26	30	38
6	02	05	09	14	16	18	20	23	27	32	36	45
7	03	05	11	16	18	21	24	26	32	37	42	53
8	03	06	12	18	21	24	27	30	36	42	48	60
9	03	07	14	20	24	27	30	34	41	47	54	68
10	04	08	15	23	26	30	34	38	45	53	60	75
11	04	08	17	25	28	33	37	41	50	58	66	83
Days.												
1	0	0	0	0	0	0	0	0	0	0	0	0
2	0	0	0	0	0	0	0	0	0	0	0	01
3	0	0	0	0	0	0	0	0	0	01	01	01
4	0	0	0	0	0	0	0	01	01	01	01	01
5	0	0	0	0	0	01	01	01	01	01	01	01
6	0	0	0	0	01	01	01	01	01	01	01	02
7	0	0	0	01	01	01	01	01	01	01	01	02
8	0	0	0	01	01	01	01	01	01	01	02	02
9	0	0	0	01	01	01	01	01	01	02	02	02
10	0	0	01	01	01	01	01	01	02	02	02	03
11	0	0	01	01	01	01	01	01	02	02	02	03
12	0	0	01	01	01	01	01	02	02	02	02	03
13	0	0	01	01	01	01	01	02	02	02	03	03
14	0	0	01	01	01	01	02	02	02	02	03	04
15	0	0	01	01	01	02	02	02	02	03	03	04
16	0	0	01	01	01	02	02	02	02	03	03	04
17	0	0	01	01	01	02	02	02	03	03	03	04
18	0	0	01	01	02	02	02	02	03	03	04	05
19	0	0	01	01	02	02	02	02	03	03	04	05
20	0	01	01	02	02	02	02	03	03	04	04	05
21	0	01	01	02	02	02	02	03	03	04	04	05
22	0	01	01	02	02	02	02	03	03	04	04	06
23	0	01	01	02	02	02	03	03	03	04	05	06
24	0	01	01	02	02	02	03	03	04	04	05	06
25	0	01	01	02	02	03	03	03	04	04	05	06
26	0	01	01	02	02	03	03	03	04	05	05	07
27	0	01	01	02	02	03	03	03	04	05	05	07
28	0	01	01	02	02	03	03	04	04	05	06	07
29	0	01	01	02	03	03	03	04	04	05	06	07
30	0	01	02	02	03	03	03	04	05	05	06	08
33	0	01	02	02	03	03	04	04	05	06	07	08
63	01	02	03	05	06	06	07	08	09	11	13	16
93	01	02	05	07	08	09	10	12	14	16	19	23

10 Dollars.

	½ per ct.	1 per ct.	2 per ct.	3 per ct.	3½ per ct.	4 per ct.	4½ per ct.	5 per ct.	6 per ct.	7 per ct.	8 per ct.	10 per ct.
Years.												
1	05	10	20	30	35	40	45	50	60	70	80	1.00
2	10	20	40	60	70	80	90	1.00	1.20	1.40	1.60	2.00
3	15	30	60	90	1.05	1.20	1.35	1.50	1.80	2.10	2.40	3.00
4	20	40	80	1.20	1.40	1.60	1.80	2.00	2.40	2.80	3.20	4.00
5	25	50	1.00	1.50	1.75	2.00	2.25	2.50	3.00	3.50	4.00	5.00
Months.												
1	0	01	02	03	03	03	04	04	05	06	07	08
2	01	02	03	05	06	07	08	08	10	12	13	17
3	01	03	05	08	09	10	11	13	15	18	20	25
4	02	03	07	10	12	13	15	17	20	23	27	33
5	02	04	08	13	15	17	19	21	25	29	33	42
6	03	05	10	15	18	20	23	25	30	35	40	50
7	03	06	12	18	20	23	26	29	35	41	47	58
8	03	07	13	20	23	27	30	33	40	47	53	67
9	04	08	15	23	26	30	34	38	45	53	60	75
10	04	08	17	25	29	33	38	42	50	58	67	83
11	05	09	18	28	32	37	41	46	55	64	73	92
Days.												
1	0	0	0	0	0	0	0	0	0	0	0	0
2	0	0	0	0	0	0	0	0	0	0	0	01
3	0	0	0	0	0	0	0	0	01	01	01	01
4	0	0	0	0	0	0	01	01	01	01	01	01
5	0	0	0	0	0	01	01	01	01	01	01	01
6	0	0	0	01	01	01	01	01	01	01	01	02
7	0	0	0	01	01	01	01	01	01	01	02	02
8	0	0	0	01	01	01	01	01	01	02	02	02
9	0	0	01	01	01	01	01	01	02	02	02	03
10	0	0	01	01	01	01	01	01	02	02	02	03
11	0	0	01	01	01	01	01	02	02	02	02	03
12	0	0	01	01	01	01	02	02	02	02	03	03
13	0	0	01	01	01	01	02	02	02	03	03	04
14	0	0	01	01	01	02	02	02	02	03	03	04
15	0	0	01	01	01	02	02	02	03	03	03	04
16	0	0	01	01	02	02	02	02	03	03	04	04
17	0	0	01	01	02	02	02	02	03	03	04	05
18	0	01	01	02	02	02	02	03	03	04	04	05
19	0	01	01	02	02	02	02	03	03	04	04	05
20	0	01	01	02	02	02	03	03	03	04	04	06
21	0	01	01	02	02	02	03	03	04	04	05	06
22	0	01	01	02	02	02	03	03	04	04	05	06
23	0	01	01	02	02	02	03	03	04	04	05	06
24	0	01	01	02	02	03	03	03	04	05	05	07
25	0	01	01	02	02	03	03	03	04	05	06	07
26	0	01	01	02	03	03	03	04	04	05	06	07
27	0	01	02	02	03	03	03	04	05	05	06	08
28	0	01	02	02	03	03	04	04	05	05	06	08
29	0	01	02	02	03	03	04	04	05	06	06	08
30	0	01	02	03	03	03	04	04	05	06	07	08
33	0	01	02	03	03	04	04	05	06	06	07	09
63	01	02	04	05	06	07	08	09	11	12	14	18
93	01	03	05	08	09	10	12	13	16	18	21	26

11 Dollars.

	¼ per ct.	1 per ct.	2 per ct.	3 per ct.	3½ per ct.	4 per ct.	4½ per ct.	5 per ct.	6 per ct.	7 per ct.	8 per ct.	10 per ct.
Years.												
1	06	11	22	33	39	44	50	55	66	77	88	1.10
2	11	22	44	66	77	88	99	1.10	1.32	1.54	1.76	2.20
3	17	33	66	99	1.16	1.32	1.49	1.65	1.98	2.31	2.64	3.30
4	22	44	88	1.32	1.54	1.76	1.98	2.20	2.64	3.08	3.52	4.40
5	28	55	1.10	1.65	1.93	2.20	2.48	2.75	3.30	3.85	4.40	5.50
Months.												
1	0	01	02	03	03	04	04	05	06	06	07	09
2	01	02	04	06	06	07	08	09	11	13	15	18
3	01	03	06	08	10	11	12	14	17	19	22	28
4	02	04	07	11	13	15	17	18	22	26	29	37
5	02	05	09	14	16	18	21	23	28	32	37	46
6	03	06	11	17	19	22	25	28	33	39	44	55
7	03	06	13	19	22	26	29	32	39	45	51	64
8	04	07	15	22	26	29	33	37	44	51	59	73
9	04	08	17	25	29	33	37	41	50	58	66	83
10	05	09	18	28	32	37	41	46	55	64	73	92
11	05	10	20	30	35	40	45	50	61	71	81	1.01
Days.												
1	0	0	0	0	0	0	0	0	0	0	0	0
2	0	0	0	0	0	0	0	0	0	0	0	01
3	0	0	0	0	0	0	0	0	01	01	01	01
4	0	0	0	0	0	0	01	01	01	01	01	01
5	0	0	0	0	01	01	01	01	01	01	01	02
6	0	0	0	01	01	01	01	01	01	01	01	02
7	0	0	0	01	01	01	01	01	01	01	02	02
8	0	0	0	01	01	01	01	01	01	02	02	02
9	0	0	01	01	01	01	01	01	02	02	02	03
10	0	0	01	01	01	01	01	02	02	02	02	03
11	0	0	01	01	01	01	02	02	02	02	03	03
12	0	0	01	01	01	01	02	02	02	03	03	04
13	0	0	01	01	01	02	02	02	02	03	03	04
14	0	0	01	01	01	02	02	02	03	03	03	04
15	0	0	01	01	02	02	02	02	03	03	04	05
16	0	0	01	01	02	02	02	02	03	03	04	05
17	0	01	01	02	02	02	02	03	03	04	04	05
18	0	01	01	02	02	02	02	03	03	04	04	06
19	0	01	01	02	02	02	03	03	03	04	05	06
20	0	01	01	02	02	02	03	03	04	04	05	06
21	0	01	01	02	02	03	03	03	04	04	05	06
22	0	01	01	02	02	03	03	03	04	05	05	07
23	0	01	01	02	02	03	03	04	04	05	06	07
24	0	01	01	02	03	03	03	04	04	05	06	07
25	0	01	02	02	03	03	03	04	05	05	06	08
26	0	01	02	02	03	03	04	04	05	06	06	08
27	0	01	02	02	03	03	04	04	05	06	07	08
28	0	01	02	03	03	03	04	04	05	06	07	09
29	0	01	02	03	03	04	04	04	05	06	07	09
30	0	01	02	03	03	04	04	05	06	06	07	09
33	01	01	02	03	04	04	05	05	06	07	08	10
63	01	02	04	06	07	08	09	10	12	13	15	19
93	01	03	06	09	10	11	13	14	17	20	23	28

12 Dollars.

Years.	½ per ct.	1 per ct.	2 per ct.	3 per ct.	3½ per ct.	4 per ct.	4½ per ct.	5 per ct.	6 per ct.	7 per ct.	8 per ct.	10 per ct.
1	06	12	24	36	42	48	54	60	72	84	96	1.20
2	12	24	48	72	84	96	1.08	1.20	1.44	1.68	1.92	2.40
3	18	36	72	1.08	1.26	1.44	1.62	1.80	2.16	2.52	2.88	3.60
4	24	48	96	1.44	1.68	1.92	2.16	2.40	2.88	3.36	3.84	4.80
5	30	60	1.20	1.80	2.10	2.40	2.70	3.00	3.60	4.20	4.80	6.00

Months.												
1	01	01	02	03	04	04	05	05	06	07	08	10
2	01	02	04	06	07	08	09	10	12	14	16	20
3	02	03	06	09	11	12	14	15	18	21	24	30
4	02	04	08	12	14	16	18	20	24	28	32	40
5	03	05	10	15	18	20	23	25	30	35	40	50
6	03	06	12	18	21	24	27	30	36	42	48	60
7	04	07	14	21	25	28	32	35	42	49	56	70
8	04	08	16	24	28	32	36	40	48	56	64	80
9	05	09	18	27	32	36	41	45	54	63	72	90
10	05	10	20	30	35	40	45	50	60	70	80	1.00
11	06	11	22	33	39	44	50	55	66	77	88	1.10

Days.												
1	0	0	0	0	0	0	0	0	0	0	0	0
2	0	0	0	0	0	0	0	0	0	0	01	01
3	0	0	0	0	0	0	0	01	01	01	01	01
4	0	0	0	0	0	01	01	01	01	01	01	01
5	0	0	0	01	01	01	01	01	01	01	01	02
6	0	0	0	01	01	01	01	01	01	01	02	02
7	0	0	0	01	01	01	01	01	01	01	02	02
8	0	0	01	01	01	01	01	01	02	02	02	03
9	0	0	01	01	01	01	01	02	02	02	02	03
10	0	0	01	01	01	01	02	02	02	02	03	03
11	0	0	01	01	01	01	02	02	02	03	03	04
12	0	0	01	01	01	02	02	02	02	03	03	04
13	0	0	01	01	02	02	02	02	03	03	03	04
14	0	0	01	01	02	02	02	02	03	03	04	05
15	0	01	01	02	02	02	02	03	03	04	04	05
16	0	01	01	02	02	02	02	03	03	04	04	05
17	0	01	01	02	02	02	03	03	03	04	05	06
18	0	01	01	02	02	02	03	03	04	04	05	06
19	0	01	01	02	02	03	03	03	04	04	05	06
20	0	01	01	02	02	03	03	03	04	05	05	07
21	0	01	01	02	02	03	03	04	04	05	06	07
22	0	01	01	02	03	03	03	04	04	05	06	07
23	0	01	02	02	03	03	03	04	05	05	06	08
24	0	01	02	02	03	03	04	04	05	06	06	08
25	0	01	02	03	03	03	04	04	05	06	07	08
26	0	01	02	03	03	03	04	04	05	06	07	09
27	0	01	02	03	03	04	04	05	05	06	07	09
28	0	01	02	03	03	04	04	05	06	07	07	09
29	0	01	02	03	03	04	04	05	06	07	08	10
30	01	01	02	03	04	04	05	05	06	07	08	10
33	01	01	02	03	04	04	05	06	07	08	09	11
63	01	02	04	06	07	08	09	11	13	15	17	21
93	02	03	06	09	11	12	14	16	19	22	25	31

13 Dollars.

Years.	½ per ct.	1 per ct.	2 per ct.	3 per ct.	3½ per ct.	4 per ct.	4½ per ct.	5 per ct.	6 per ct.	7 per ct.	8 per ct.	10 per ct.
1	07	13	26	39	46	52	59	65	78	91	1.04	1.30
2	13	26	52	78	91	1.04	1.17	1.30	1.56	1.82	2.08	2.60
3	20	39	78	1.17	1.37	1.56	1.76	1.95	2.34	2.73	3.12	3.90
4	26	52	1.04	1.56	1.82	2.08	2.34	2.60	3.12	3.64	4.16	5.20
5	33	65	1.30	1.95	2.28	2.60	2.93	3.25	3.90	4.55	5.20	6.50
Months.												
1	01	01	02	03	04	04	05	05	07	08	09	11
2	01	02	04	07	08	09	10	11	13	15	17	22
3	02	03	07	10	11	13	15	16	20	23	26	33
4	02	04	09	13	15	17	20	22	26	30	35	43
5	03	05	11	16	19	22	24	27	33	38	43	54
6	03	07	13	20	23	26	29	33	39	46	52	65
7	04	08	15	23	27	30	34	38	46	53	61	76
8	04	09	17	26	30	35	39	43	52	61	69	87
9	05	10	20	29	34	39	44	49	59	68	78	98
10	05	11	22	33	38	43	49	54	65	76	87	1.08
11	06	12	24	36	42	48	54	60	72	83	95	1.19
Days.												
1	0	0	0	0	0	0	0	0	0	0	0	0
2	0	0	0	0	0	0	0	0	0	0	01	01
3	0	0	0	0	0	0	0	01	01	01	01	01
4	0	0	0	0	01	01	01	01	01	01	01	01
5	0	0	0	01	01	01	01	01	01	01	01	02
6	0	0	0	01	01	01	01	01	01	02	02	02
7	0	0	01	01	01	01	01	01	02	02	02	03
8	0	0	01	01	01	01	01	01	02	02	02	03
9	0	0	01	01	01	01	01	02	02	02	03	03
10	0	0	01	01	01	01	02	02	02	03	03	04
11	0	0	01	01	01	02	02	02	02	03	03	04
12	0	0	01	01	02	02	02	02	03	03	03	04
13	0	0	01	01	02	02	02	02	03	03	04	05
14	0	01	01	02	02	02	02	03	03	04	04	05
15	0	01	01	02	02	02	02	03	03	04	04	05
16	0	01	01	02	02	02	03	03	03	04	05	06
17	0	01	01	02	02	02	03	03	04	04	05	06
18	0	01	01	02	02	03	03	03	04	05	05	07
19	0	01	01	02	02	03	03	03	04	05	05	07
20	0	01	01	02	03	03	03	04	04	05	06	07
21	0	01	02	02	03	03	03	04	05	05	06	08
22	0	01	02	02	03	03	04	04	05	06	06	08
23	0	01	02	02	03	03	04	04	05	06	07	08
24	0	01	02	03	03	03	04	04	05	06	07	09
25	0	01	02	03	03	04	04	05	05	06	07	09
26	0	01	02	03	03	04	04	05	06	07	08	09
27	0	01	02	03	03	04	04	05	06	07	08	10
28	01	01	02	03	04	04	05	05	06	07	08	10
29	01	01	02	03	04	04	05	05	06	07	08	10
30	01	01	02	03	04	04	05	05	07	08	09	11
33	01	01	02	04	04	05	05	06	07	08	10	12
63	01	02	05	07	08	09	10	11	14	16	18	23
93	02	03	07	10	12	13	15	17	20	24	27	34

14 Dollars.

Years.	½ per ct.	1 per ct.	2 per ct.	3 per ct.	3½ per ct.	4 per ct.	4½ per ct.	5 per ct.	6 per ct.	7 per ct.	8 per ct.	10 per ct.
1	07	14	28	42	49	56	63	70	84	98	1.12	1.40
2	14	28	56	84	98	1.12	1.26	1.40	1.68	1.96	2.24	2.80
3	21	42	84	1.26	1.47	1.68	1.89	2.10	2.52	2.94	3.36	4.20
4	28	56	1.12	1.68	1.96	2.24	2.52	2.80	3.36	3.92	4.48	5.60
5	35	70	1.40	2.10	2.45	2.80	3.15	3.50	4.20	4.90	5.60	7.00
Months.												
1	01	01	02	04	04	05	05	06	07	08	09	12
2	01	02	05	07	08	09	11	12	14	16	19	23
3	02	04	07	11	12	14	16	18	21	25	28	35
4	02	05	09	14	16	19	21	23	28	33	37	47
5	03	06	12	18	20	23	26	29	35	41	47	58
6	04	07	14	21	25	28	32	35	42	49	56	70
7	04	08	16	25	29	33	37	41	49	57	65	82
8	05	09	19	28	33	37	42	47	55	65	75	93
9	05	11	21	32	37	42	47	53	63	74	84	1.05
10	06	12	23	35	41	47	53	58	70	82	93	1.17
11	06	13	26	39	45	51	58	64	77	90	1.03	1.28
Days.												
1	0	0	0	0	0	0	0	0	0	0	0	0
2	0	0	0	0	0	0	0	0	0	01	01	01
3	0	0	0	0	0	0	01	01	01	01	01	01
4	0	0	0	0	01	01	01	01	01	01	01	02
5	0	0	0	01	01	01	01	01	01	01	02	02
6	0	0	0	01	01	01	01	01	01	01	02	02
7	0	0	01	01	01	01	01	01	02	02	02	03
8	0	0	01	01	01	01	01	02	02	02	02	03
9	0	0	01	01	01	01	02	02	02	02	03	04
10	0	0	01	01	01	02	02	02	02	03	03	04
11	0	0	01	01	01	02	02	02	03	03	03	04
12	0	0	01	01	02	02	02	02	03	03	04	05
13	0	01	01	02	02	02	02	03	03	04	04	05
14	0	01	01	02	02	02	02	03	03	04	04	05
15	0	01	01	02	02	02	03	03	04	04	05	06
16	0	01	01	02	02	02	03	03	04	04	05	06
17	0	01	01	02	02	03	03	03	04	05	05	07
18	0	01	01	02	02	03	03	04	04	05	06	07
19	0	01	01	02	03	03	03	04	04	05	06	07
20	0	01	02	02	03	03	04	04	05	05	06	08
21	0	01	02	02	03	03	04	04	05	06	07	08
22	0	01	02	03	03	03	04	04	05	06	07	09
23	0	01	02	03	03	04	04	04	05	06	07	09
24	0	01	02	03	03	04	04	05	06	07	07	09
25	0	01	02	03	03	04	04	05	06	07	08	10
26	01	01	02	03	04	04	05	05	06	07	08	10
27	01	01	02	03	04	04	05	05	06	07	08	11
28	01	01	02	03	04	04	05	05	07	08	09	11
29	01	01	02	03	04	05	05	06	07	08	09	11
30	01	01	02	04	04	05	05	06	07	08	09	12
33	01	01	03	04	04	05	06	06	08	09	10	13
63	01	02	05	07	09	09	11	12	15	17	20	25
93	02	04	07	11	13	14	16	18	22	25	29	36

15 Dollars.

Years.	½ per ct.	1 per ct.	2 per ct.	3 per ct.	3½ per ct.	4 per ct.	4½ per ct.	5 per ct.	6 per ct.	7 per ct.	8 per ct.	10 per ct.
1	08	15	30	45	53	60	68	75	90	1.05	1.20	1.50
2	15	30	60	90	1.05	1.20	1.35	1.50	1.80	2.10	2.40	3.00
3	23	45	90	1.35	1.58	1.80	2.03	2.25	2.70	3.15	3.60	4.50
4	30	60	1.20	1.80	2.10	2.40	2.70	3.00	3.60	4.20	4.80	6.00
5	38	75	1.50	2.25	2.63	3.00	3.38	3.75	4.50	5.25	6.00	7.50
Months.												
1	01	01	03	04	04	05	06	06	08	09	10	13
2	01	03	05	08	09	10	11	13	15	18	20	25
3	02	04	08	11	13	15	17	19	23	26	30	38
4	03	05	10	15	18	20	23	25	30	35	40	50
5	03	06	13	19	22	25	28	31	38	44	50	63
6	04	08	15	23	26	30	34	38	45	53	60	75
7	04	09	18	26	31	35	39	44	53	61	70	88
8	05	10	20	30	35	40	45	50	60	70	80	1.00
9	06	11	23	34	39	45	51	56	68	79	90	1.13
10	06	13	25	38	44	50	56	63	75	88	1.00	1.25
11	07	14	28	41	48	55	62	69	83	96	1.10	1.38
Days.												
1	0	0	0	0	0	0	0	0	0	0	0	0
2	0	0	0	0	0	0	0	0	01	01	01	01
3	0	0	0	0	0	01	01	01	01	01	01	01
4	0	0	0	01	01	01	01	01	01	01	01	02
5	0	0	0	01	01	01	01	01	01	01	02	02
6	0	0	01	01	01	01	01	01	02	02	02	03
7	0	0	01	01	01	01	01	01	02	02	02	03
8	0	0	01	01	01	01	02	02	02	02	03	03
9	0	0	01	01	01	02	02	02	02	03	03	04
10	0	0	01	01	01	02	02	02	03	03	03	04
11	0	0	01	01	02	02	02	02	03	03	04	05
12	0	01	01	02	02	02	02	03	03	04	04	05
13	0	01	01	02	02	02	02	03	03	04	04	05
14	0	01	01	02	02	02	03	03	04	04	05	06
15	0	01	01	02	02	03	03	03	04	04	05	06
16	0	01	01	02	02	03	03	03	04	05	05	07
17	0	01	01	02	02	03	03	04	04	05	06	07
18	0	01	02	02	03	03	03	04	05	05	06	08
19	0	01	02	02	03	03	04	04	05	06	06	08
20	0	01	02	03	03	03	04	04	05	06	07	08
21	0	01	02	03	03	04	04	04	05	06	07	09
22	0	01	02	03	03	04	04	05	06	06	07	09
23	0	01	02	03	03	04	04	05	06	07	08	10
24	01	01	02	03	04	04	05	05	06	07	08	10
25	01	01	02	03	04	04	05	05	06	07	08	10
26	01	01	02	03	04	04	05	05	07	08	09	11
27	01	01	02	03	04	05	05	06	07	08	09	11
28	01	01	02	04	04	05	05	06	07	08	09	12
29	01	01	02	04	04	05	05	06	07	08	10	12
30	01	01	03	04	04	05	06	06	08	09	10	13
33	01	01	03	04	05	06	06	07	08	10	11	14
63	01	03	05	08	09	11	12	13	16	18	21	26
93	02	04	08	12	14	16	17	19	23	27	31	39

16 Dollars.

	½ per ct.	1 per ct.	2 per ct.	3 per ct.	3½ per ct.	4 per ct.	4½ per ct.	5 per ct.	6 per ct.	7 per ct.	8 per ct.	10 per ct.
Years.												
1	08	16	32	48	54	64	72	80	96	1.12	1.28	1.60
2	16	32	64	96	1.08	1.28	1.44	1.60	1.92	2.24	2.56	3.20
3	24	48	96	1.44	1.62	1.92	2.16	2.40	2.88	3.36	3.84	4.80
4	32	64	1.28	1.92	2.16	2.56	2.88	3.20	3.84	4.48	5.12	6.40
5	40	80	1.60	2.40	2.70	3.20	3.60	4.00	4.80	5.60	6.40	8.00
Months.												
1	01	01	03	04	05	05	06	07	08	09	11	13
2	01	03	05	08	09	11	12	13	16	19	21	27
3	02	04	08	12	14	16	18	20	24	28	32	40
4	03	05	11	16	18	21	24	27	32	37	43	53
5	03	07	13	20	23	27	30	33	40	47	53	67
6	04	08	16	24	27	32	36	40	48	56	64	80
7	05	09	19	28	32	37	42	47	56	65	75	93
8	05	11	21	32	36	43	48	53	64	75	85	1.07
9	06	12	24	36	41	48	54	60	72	84	96	1.20
10	07	13	27	40	45	53	60	67	80	93	1.07	1.33
11	07	15	29	44	50	59	66	73	88	1.03	1.17	1.47
Days.												
1	0	0	0	0	0	0	0	0	0	0	0	0
2	0	0	0	0	0	0	0	0	01	01	01	01
3	0	0	0	0	0	01	01	01	01	01	01	01
4	0	0	0	01	01	01	01	01	01	01	01	02
5	0	0	0	01	01	01	01	01	01	02	02	02
6	0	0	01	01	01	01	01	01	02	02	02	03
7	0	0	01	01	01	01	01	02	02	02	02	03
8	0	0	01	01	01	01	02	02	02	02	03	04
9	0	0	01	01	01	02	02	02	02	02	03	04
10	0	0	01	01	02	02	02	02	03	03	04	04
11	0	0	01	01	02	02	02	02	03	03	04	05
12	0	01	01	02	02	02	02	03	03	04	04	05
13	0	01	01	02	02	02	03	03	03	04	05	06
14	0	01	01	02	02	02	03	03	04	04	05	06
15	0	01	01	02	02	03	03	03	04	05	05	07
16	0	01	01	02	02	03	03	04	04	05	06	07
17	0	01	02	02	03	03	03	04	05	05	06	08
18	0	01	02	02	03	03	04	04	05	06	06	08
19	0	01	02	03	03	03	04	04	05	06	07	08
20	0	01	02	03	03	04	04	04	05	06	07	09
21	0	01	02	03	03	04	04	05	06	07	07	09
22	0	01	02	03	03	04	04	05	06	07	08	10
23	01	01	02	03	03	04	05	05	06	07	08	10
24	01	01	02	03	04	04	05	05	06	07	09	11
25	01	01	02	03	04	04	05	06	07	08	09	11
26	01	01	02	03	04	05	05	06	07	08	09	12
27	01	01	02	04	04	05	05	06	07	08	10	12
28	01	01	02	04	04	05	06	06	07	09	10	12
29	01	01	03	04	04	05	06	06	08	09	10	13
30	01	01	03	04	05	05	06	07	08	09	11	13
33	01	01	03	04	05	06	07	07	09	10	12	15
63	01	03	06	08	10	11	13	14	17	20	22	28
93	02	04	08	12	14	17	19	21	25	29	33	41

17 Dollars.

Years.	½ per ct.	1 per ct.	2 per ct.	3 per ct.	3½ per ct.	4 per ct.	4½ per ct.	5 per ct.	6 per ct.	7 per ct.	8 per ct.	10 per ct.
1	09	17	34	51	60	68	77	85	1.02	1.19	1.36	1.70
2	17	34	68	1.02	1.19	1.36	1.53	1.70	2.04	2.38	2.72	3.40
3	26	51	1.02	1.53	1.79	2.04	2.30	2.55	3.06	3.57	4.08	5.10
4	34	68	1.36	2.04	2.38	2.72	3.06	3.40	4.08	4.76	5.44	6.80
5	43	85	1.70	2.55	2.98	3.40	3.83	4.25	5.10	5.95	6.80	8.50
Months.												
1	01	01	03	04	05	06	06	07	09	10	11	14
2	01	03	06	09	10	11	13	14	17	20	23	28
3	02	04	09	13	15	17	19	21	26	30	34	43
4	03	06	11	17	20	23	26	28	34	40	45	57
5	04	07	14	21	25	28	32	35	43	50	57	71
6	04	09	17	26	30	34	38	43	51	60	68	85
7	05	10	20	30	35	40	45	50	60	69	79	99
8	06	11	23	34	40	45	51	57	68	79	91	1.13
9	06	13	26	38	45	51	57	64	77	89	1.02	1.28
10	07	14	28	43	50	57	64	71	85	99	1.13	1.42
11	08	16	31	47	55	62	70	78	94	1.09	1.25	1.56
Days.												
1	0	0	0	0	0	0	0	0	0	0	0	0
2	0	0	0	0	0	0	0	0	01	01	01	01
3	0	0	0	0	01	01	01	01	01	01	01	01
4	0	0	0	01	01	01	01	01	01	01	02	02
5	0	0	0	01	01	01	01	01	01	02	02	02
6	0	0	01	01	01	01	01	01	02	02	02	03
7	0	0	01	01	01	01	01	02	02	02	02	03
8	0	0	01	01	01	02	02	02	02	03	03	04
9	0	0	01	01	01	02	02	02	03	03	03	04
10	0	0	01	01	02	02	02	02	03	03	04	05
11	0	01	01	02	02	02	02	03	03	04	04	05
12	0	01	01	02	02	02	03	03	03	04	05	06
13	0	01	01	02	02	02	03	03	04	04	05	06
14	0	01	01	02	02	03	03	03	04	05	05	07
15	0	01	01	02	02	03	03	04	04	05	06	07
16	0	01	02	02	03	03	03	04	05	05	06	08
17	0	01	02	02	03	03	04	04	05	06	06	08
18	0	01	02	03	03	03	04	04	05	06	07	09
19	0	01	02	03	03	04	04	04	05	06	07	09
20	0	01	02	03	03	04	04	05	06	07	08	09
21	0	01	02	03	03	04	04	05	06	07	08	10
22	01	01	02	03	04	04	05	05	06	07	08	10
23	01	01	02	03	04	04	05	05	07	08	09	11
24	01	01	02	03	04	05	05	06	07	08	09	11
25	01	01	02	04	04	05	05	06	07	08	09	12
26	01	01	02	04	04	05	06	06	07	09	10	12
27	01	01	03	04	04	05	06	06	08	09	10	13
28	01	01	03	04	05	05	06	07	08	09	11	13
29	01	01	03	04	05	05	06	07	08	10	11	14
30	01	01	03	04	05	06	06	07	09	10	11	14
33	01	02	03	05	05	06	07	08	09	11	12	16
63	01	03	06	09	10	12	13	15	18	21	24	30
93	02	04	09	13	15	18	20	22	26	31	35	44

18 Dollars.

Years.	½ per ct.	1 per ct.	2 per ct.	3 per ct.	3½ per ct.	4 per ct.	4½ per ct.	5 per ct.	6 per ct.	7 per ct.	8 per ct.	10 per ct.
1	09	18	36	54	63	72	81	90	1.08	1.26	1.44	1.80
2	18	36	72	1.08	1.26	1.44	1.62	1.80	2.16	2.52	2.88	3.60
3	27	54	1.08	1.62	1.89	2.16	2.43	2.70	3.24	3.78	4.32	5.40
4	36	72	1.44	2.16	2.52	2.88	3.24	3.60	4.32	5.04	5.76	7.20
5	45	90	1.80	2.70	3.15	3.60	4.05	4.50	5.40	6.30	7.20	9.00
Months.												
1	01	02	03	05	05	06	07	08	09	11	12	15
2	02	03	06	09	11	12	14	15	18	21	24	30
3	02	05	09	14	16	18	20	23	27	32	36	45
4	03	06	12	18	21	24	27	30	36	42	48	60
5	04	08	15	23	26	30	34	38	45	53	60	75
6	05	09	18	27	32	36	41	45	54	63	72	90
7	05	11	21	32	37	42	47	53	63	74	84	1.05
8	06	12	24	36	42	48	54	60	72	84	96	1.20
9	07	14	27	41	47	54	61	68	81	95	1.08	1.35
10	08	15	30	45	53	60	68	75	90	1.05	1.20	1.50
11	08	17	33	50	58	66	74	83	99	1.16	1.32	1.65
Days.												
1	0	0	0	0	0	0	0	0	0	0	0	01
2	0	0	0	0	0	0	0	01	01	01	01	01
3	0	0	0	0	01	01	01	01	01	01	01	02
4	0	0	0	01	01	01	01	01	01	01	02	02
5	0	0	01	01	01	01	01	01	02	02	02	03
6	0	0	01	01	01	01	01	02	02	02	02	03
7	0	0	01	01	01	01	02	02	02	02	03	04
8	0	0	01	01	01	02	02	02	02	03	03	04
9	0	0	01	01	02	02	02	02	03	03	04	05
10	0	01	01	02	02	02	02	03	03	04	04	05
11	0	01	01	02	02	02	02	03	03	04	04	06
12	0	01	01	02	02	02	03	03	04	04	05	06
13	0	01	01	02	02	03	03	03	04	05	05	07
14	0	01	01	02	02	03	03	04	04	05	06	07
15	0	01	02	02	03	03	03	04	05	05	06	08
16	0	01	02	02	03	03	04	04	05	06	06	08
17	0	01	02	03	03	03	04	04	05	06	07	09
18	0	01	02	03	03	04	04	05	05	06	07	09
19	0	01	02	03	03	04	04	05	06	07	08	10
20	01	01	02	03	04	04	05	05	06	07	08	10
21	01	01	02	03	04	04	05	05	06	07	08	11
22	01	01	02	03	04	04	05	06	07	08	09	11
23	01	01	02	03	04	05	05	06	07	08	09	12
24	01	01	02	04	04	05	05	06	07	08	10	12
25	01	01	03	04	04	05	06	06	08	09	10	13
26	01	01	03	04	05	05	06	07	08	09	10	13
27	01	01	03	04	05	05	06	07	08	09	11	14
28	01	01	03	04	05	06	06	07	08	10	11	14
29	01	01	03	04	05	06	07	07	09	10	12	15
30	01	02	03	05	05	06	07	08	09	11	12	15
33	01	02	03	05	06	07	07	08	10	12	13	17
63	02	03	06	09	11	13	14	16	19	22	25	32
93	02	05	09	14	16	19	21	23	28	33	37	47

19 Dollars.

Years.	½ per ct.	1 per ct.	2 per et.	3 per ct.	3½ per ct.	4 per ct.	4½ per ct.	5 per ct.	6 per ct.	7 per ct.	8 per ct.	10 per ct.
1	10	19	38	57	67	76	86	95	1.14	1.33	1.52	1.90
2	19	38	76	1.14	1.33	1.52	1.71	1.90	2.28	2.66	3.04	3.80
3	29	57	1.14	1.71	2.00	2.28	2.57	2.85	3.42	3.99	4.56	5.70
4	38	76	1.52	2.28	2.66	3.04	3.42	3.80	4.56	5.32	6.08	7.60
5	48	95	1.90	2.85	3.33	3.80	4.28	4.75	5.70	6.65	7.60	9.50
Months.												
1	01	02	03	05	06	06	07	08	10	11	13	16
2	02	03	06	10	11	13	14	16	19	22	25	32
3	02	05	10	14	17	19	21	24	29	33	38	48
4	03	06	13	19	22	25	29	32	38	44	51	63
5	04	08	16	24	28	32	36	40	48	55	63	79
6	05	10	19	29	33	38	43	48	57	67	76	95
7	06	11	22	33	39	44	50	55	67	78	89	1.11
8	06	13	25	38	44	51	57	63	76	89	1.01	1.27
9	07	14	29	43	50	57	64	71	86	1.00	1.14	1.43
10	08	16	32	48	55	63	71	79	95	1.11	1.27	1.58
11	09	17	35	52	61	70	78	87	1.05	1.22	1.39	1.74
Days.												
1	0	0	0	0	0	0	0	0	0	0	0	01
2	0	0	0	0	0	0	0	01	01	01	01	01
3	0	0	0	0	01	01	01	01	01	01	01	02
4	0	0	0	01	01	01	01	01	01	01	02	02
5	0	0	01	01	01	01	01	01	02	02	02	03
6	0	0	01	01	01	01	01	02	02	02	03	03
7	0	0	01	01	01	01	02	02	02	03	03	04
8	0	0	01	01	01	02	02	02	03	03	03	04
9	0	0	01	01	02	02	02	02	03	03	04	05
10	0	01	01	02	02	02	02	03	03	04	04	05
11	0	01	01	02	02	02	03	03	03	04	05	06
12	0	01	01	02	02	03	03	03	04	04	05	06
13	0	01	01	02	02	03	03	03	04	05	05	07
14	0	01	01	02	03	03	03	04	04	05	06	07
15	0	01	02	02	03	03	04	04	05	06	06	08
16	0	01	02	03	03	03	04	04	05	06	07	08
17	0	01	02	03	03	04	04	04	05	06	07	09
18	0	01	02	03	03	04	04	05	06	07	08	10
19	01	01	02	03	04	04	05	05	06	07	08	10
20	01	01	02	03	04	04	05	05	06	07	08	11
21	01	01	02	03	04	04	05	06	07	08	09	11
22	01	01	02	03	04	05	05	06	07	08	09	12
23	01	01	02	04	04	05	05	06	07	08	10	12
24	01	01	03	04	04	05	06	06	08	09	10	13
25	01	01	03	04	05	05	06	07	08	09	11	13
26	01	01	03	04	05	05	06	07	08	10	11	14
27	01	01	03	04	05	06	06	07	09	10	11	14
28	01	01	03	04	05	06	07	07	09	10	12	15
29	01	02	03	05	05	06	07	08	09	11	12	15
30	01	02	03	05	06	06	07	08	10	11	13	16
33	01	02	03	05	06	07	08	09	10	12	14	17
63	02	03	07	10	12	13	15	17	20	23	27	33
93	02	05	10	15	17	20	22	25	29	34	39	49

20 Dollars.

Years.	½ per ct.	1 per ct.	2 per ct.	3 per ct.	3½ per ct.	4 per ct.	4½ per ct.	5 per ct.	6 per ct.	7 per ct.	8 per ct.	10 per ct.
1	10	20	40	60	70	80	90	1.00	1.20	1.40	1.60	2.00
2	20	40	80	1.20	1.40	1.60	1.80	2.00	2.40	2.80	3.20	4.00
3	30	60	1.20	1.80	2.10	2.40	2.70	3.00	3.60	4.20	4.80	6.00
4	40	80	1.60	2.40	2.80	3.20	3.60	4.00	4.80	5.60	6.40	8.00
5	50	1.00	2.00	3.00	3.50	4.00	4.50	5.00	6.00	7.00	8.00	10.00
Months.												
1	01	02	03	05	06	07	08	08	10	12	13	17
2	02	03	07	10	12	13	15	17	20	23	27	33
3	03	05	10	15	18	20	23	25	30	35	40	50
4	03	07	13	20	23	27	30	33	40	47	53	67
5	04	08	17	25	29	33	38	42	50	58	67	83
6	05	10	20	30	35	40	45	50	60	70	80	1.00
7	06	12	23	35	41	47	53	58	70	82	93	1.17
8	07	13	27	40	47	53	60	67	80	93	1.07	1.33
9	08	15	30	45	53	60	68	75	90	1.05	1.20	1.50
10	08	17	33	50	58	67	75	83	1.00	1.17	1.33	1.67
11	09	18	37	55	64	73	83	92	1.10	1.28	1.47	1.83
Days.												
1	0	0	0	0	0	0	0	0	0	0	0	01
2	0	0	0	0	0	0	01	01	01	01	01	01
3	0	0	0	01	01	01	01	01	01	01	01	02
4	0	0	0	01	01	01	01	01	01	02	02	02
5	0	0	01	01	01	01	01	01	02	02	02	03
6	0	0	01	01	01	01	02	02	02	02	03	03
7	0	0	01	01	01	02	02	02	02	03	03	04
8	0	0	01	01	02	02	02	02	03	03	04	04
9	0	01	01	02	02	02	02	03	03	04	04	05
10	0	01	01	02	02	02	03	03	03	04	04	06
11	0	01	01	02	02	02	03	03	04	04	05	06
12	0	01	01	02	02	03	03	03	04	05	05	07
13	0	01	01	02	03	03	03	04	04	05	06	07
14	0	01	02	02	03	03	04	04	05	05	06	08
15	0	01	02	03	03	03	04	04	05	06	07	08
16	0	01	02	03	03	04	04	04	05	06	07	09
17	0	01	02	03	03	04	04	05	06	07	08	09
18	01	01	02	03	04	04	05	05	06	07	08	10
19	01	01	02	03	04	04	05	05	06	07	08	11
20	01	01	02	03	04	04	05	06	07	08	09	11
21	01	01	02	04	04	05	05	06	07	08	09	12
22	01	01	02	04	04	05	06	06	07	09	10	12
23	01	01	03	04	04	05	06	06	08	09	10	13
24	01	01	03	04	05	05	06	07	08	09	11	13
25	01	01	03	04	05	06	06	07	08	10	11	14
26	01	01	03	04	05	06	07	07	09	10	12	14
27	01	02	03	05	05	06	07	08	09	11	12	15
28	01	02	03	05	05	06	07	08	09	11	12	16
29	01	02	03	05	06	06	07	08	10	11	13	16
30	01	02	03	05	06	07	08	08	10	12	13	17
33	01	02	04	06	06	07	08	09	11	13	15	18
63	02	04	07	11	12	14	16	18	21	25	28	35
93	03	05	10	16	18	21	23	26	31	36	41	52

21 Dollars.

	½ per ct.	1 per ct.	2 per ct.	3 per ct.	3½ per ct.	4 per ct.	4½ per ct.	5 per ct.	6 per ct.	7 per ct.	8 per ct.	10 per ct.
Years.												
1	11	21	42	63	74	84	95	1.05	1.26	1.47	1.68	2.10
2	21	42	84	1.26	1.47	1.68	1.89	2.10	2.52	2.94	3.36	4.20
3	32	63	1.26	1.89	2.21	2.52	2.84	3.15	3.78	4.41	5.04	6.30
4	42	84	1.68	2.52	2.94	3.36	3.78	4.20	5.04	5.88	6.72	8.40
5	53	1.05	2.10	3.15	3.68	4.20	4.73	5.25	6.30	7.35	8.40	10.50
Months.												
1	01	02	04	05	06	07	08	09	11	12	14	18
2	02	04	07	11	12	14	16	18	21	25	28	35
3	03	05	11	16	18	21	24	26	32	37	42	53
4	04	07	14	21	25	28	32	35	42	49	56	70
5	04	09	18	26	31	35	39	44	53	61	70	88
6	05	11	21	32	37	42	47	53	63	74	84	1.05
7	06	12	25	37	43	49	55	61	74	86	98	1.23
8	07	14	28	42	49	56	63	70	84	98	1.12	1.40
9	08	16	32	47	55	63	71	79	95	1.10	1.26	1.58
10	09	18	35	53	61	70	79	88	1.05	1.23	1.40	1.75
11	10	19	39	58	67	77	87	96	1.16	1.35	1.54	1.93
Days.												
1	0	0	0	0	0	0	0	0	0	0	0	01
2	0	0	0	0	0	0	01	01	01	01	01	01
3	0	0	0	01	01	01	01	01	01	01	01	02
4	0	0	0	01	01	01	01	01	01	02	02	02
5	0	0	01	01	01	01	01	01	02	02	02	03
6	0	0	01	01	01	01	02	02	02	02	03	04
7	0	0	01	01	01	02	02	02	02	03	03	04
8	0	0	01	01	02	02	02	02	03	03	04	05
9	0	01	01	02	02	02	02	03	03	04	04	05
10	0	01	01	02	02	02	03	03	04	04	05	06
11	0	01	01	02	02	03	03	03	04	04	05	06
12	0	01	01	02	02	03	03	04	04	05	06	07
13	0	01	02	02	03	03	03	04	05	05	06	08
14	0	01	02	02	03	03	04	04	05	06	07	08
15	0	01	02	03	03	04	04	04	05	06	07	09
16	0	01	02	03	03	04	04	05	06	07	07	09
17	0	01	02	03	03	04	04	05	06	07	08	10
18	01	01	02	03	04	04	05	05	06	07	08	11
19	01	01	02	03	04	04	05	06	07	08	09	11
20	01	01	02	04	04	05	05	06	07	08	09	12
21	01	01	02	04	04	05	06	06	07	09	10	12
22	01	01	03	04	04	05	06	06	08	09	10	13
23	01	01	03	04	05	05	06	07	08	09	11	13
24	01	01	03	04	05	06	06	07	08	10	11	14
25	01	01	03	04	05	06	07	07	09	10	12	15
26	01	02	03	05	05	06	07	08	09	11	12	15
27	01	02	03	05	06	06	07	08	09	11	13	16
28	01	02	03	05	06	07	07	08	10	11	13	16
29	01	02	03	05	06	07	08	08	10	12	14	17
30	01	02	04	05	06	07	08	09	11	12	14	18
33	01	02	04	06	07	08	09	10	12	13	15	19
63	02	04	07	11	13	15	17	18	22	26	29	37
93	03	05	11	16	19	22	24	27	33	38	43	54

22 Dollars.

Years.	½ per ct.	1 per ct.	2 per ct.	3 per ct.	3½ per ct.	4 per ct.	4½ per ct.	5 per ct.	6 per ct.	7 per ct.	8 per ct.	10 per ct.
1	11	22	44	66	77	88	99	1.10	1.32	1.54	1.76	2.20
2	22	44	88	1.32	1.54	1.76	1.98	2.20	2.64	3.08	3.52	4.40
3	33	66	1.32	1.98	2.31	2.64	2.97	3.30	3.96	4.62	5.28	6.60
4	44	88	1.76	2.64	3.08	3.52	3.96	4.40	5.28	6.16	7.04	8.80
5	55	1.10	2.20	3.30	3.85	4.40	4.95	5.50	6.60	7.70	8.80	11.00
Months.												
1	01	02	04	06	06	07	08	09	11	13	15	18
2	02	04	07	11	13	15	17	18	22	26	29	37
3	03	06	11	17	19	22	25	28	33	39	44	55
4	04	07	15	22	26	29	33	46	55	64	73	92
5	05	09	18	28	32	37	41	46	55	64	73	92
6	06	11	22	33	39	44	50	55	66	77	88	1.10
7	06	13	26	39	45	51	58	64	77	90	1.03	1.28
8	07	15	29	44	51	59	66	73	88	1.03	1.17	1.47
9	08	17	33	50	58	66	74	83	99	1.16	1.32	1.65
10	09	18	37	55	64	73	83	92	1.10	1.28	1.47	1.83
11	10	20	40	61	71	81	91	1.01	1.21	1.41	1.61	2.02
Days.												
1	0	0	0	0	0	0	0	0	0	0	0	01
2	0	0	0	0	0	0	01	01	01	01	01	01
3	0	0	0	01	01	01	01	01	01	01	01	02
4	0	0	0	01	01	01	01	01	01	02	02	02
5	0	0	01	01	01	01	01	02	02	02	02	03
6	0	0	01	01	01	01	02	02	02	03	03	04
7	0	0	01	01	01	02	02	02	03	03	03	04
8	0	0	01	01	02	02	02	02	03	03	04	05
9	0	01	01	02	02	02	02	03	03	04	04	06
10	0	01	01	02	02	02	03	03	04	04	05	06
11	0	01	01	02	02	03	03	03	04	05	05	07
12	0	01	01	02	03	03	03	04	04	05	06	07
13	0	01	02	02	03	03	04	04	05	06	06	08
14	0	01	02	03	03	03	04	04	05	06	07	09
15	0	01	02	03	03	04	04	05	06	06	07	09
16	0	01	02	03	03	04	04	05	06	07	08	10
17	01	01	02	03	04	04	05	05	06	07	08	10
18	01	01	02	03	04	04	05	06	07	08	09	11
19	01	01	02	03	04	05	05	06	07	08	09	12
20	01	01	02	04	04	05	06	06	07	09	10	12
21	01	01	03	04	04	05	06	06	08	09	10	13
22	01	01	03	04	05	05	06	07	08	09	11	13
23	01	01	03	04	05	06	06	07	08	10	11	14
24	01	01	03	04	05	06	07	07	09	10	12	15
25	01	02	03	05	05	06	07	08	09	11	12	15
26	01	02	03	05	06	06	07	08	10	11	13	16
27	01	02	03	05	06	07	07	08	10	12	13	17
28	01	02	03	05	06	07	08	09	10	12	14	17
29	01	02	04	05	06	07	08	09	11	12	14	18
30	01	02	04	06	06	07	08	09	11	13	15	18
33	01	02	04	06	07	08	09	10	12	14	16	20
63	02	04	08	12	13	15	17	19	23	27	31	39
93	03	06	11	17	20	23	26	28	34	40	45	57

23 Dollars.

	½ per ct.	1 per ct.	2 per ct.	3 per ct.	3½ per ct.	4 per ct.	4½ per ct.	5 per ct.	6 per ct.	7 per ct.	8 per ct.	10 per ct.
Years.												
1	12	23	46	69		81	92	1.04	1.15	1.38	1.61	1.84	2.30
2	23	46	92	1.38	1.61	1.84	2.07	2.30	2.76	3.22	3.68	4.60
3	35	69	1.38	2.07	2.42	2.76	3.11	3.45	4.14	4.83	5.52	6.90
4	46	92	1.84	2.76	3.22	3.68	4.14	4.60	5.52	6.44	7.36	9.20
5	58	1.15	2.30	3.45	4.03	4.60	5.18	5.75	6.90	8.05	9.20	11.50
Months.												
1	01	02	04	06	07	08	09	10	12	13	15	19
2	02	04	08	12	13	15	17	19	23	27	31	38
3	03	06	12	17	20	23	26	29	35	40	46	58
4	04	08	15	23	27	31	35	38	46	54	61	77
5	05	10	19	29	34	38	43	48	58	67	77	96
6	06	12	23	35	40	46	52	58	69	81	92	1.15
7	07	13	27	40	47	54	60	67	81	94	1.07	1.34
8	08	15	31	46	54	61	69	77	92	1.07	1.23	1.53
9	09	17	35	52	60	69	78	86	1.04	1.21	1.38	1.73
10	10	19	38	58	67	77	86	96	1.15	1.34	1.53	1.92
11	11	21	42	63	74	84	95	1.05	1.27	1.48	1.69	2.11
Days.												
1	0	0	0	0	0	0	0	0	0	0	01	01
2	0	0	0	0	0	01	01	01	01	01	01	01
3	0	0	0	01	01	01	01	01	01	01	02	02
4	0	0	01	01	01	01	01	01	01	01	02	03
5	0	0	01	01	01	01	01	02	02	02	03	03
6	0	0	01	01	01	02	02	02	02	02	03	03
7	0	0	01	01	02	02	02	02	03	03	03	04
8	0	0	01	02	02	02	02	03	03	04	04	04
9	0	01	01	02	02	02	03	03	03	04	05	05
10	0	01	01	02	02	03	03	03	04	04	05	06
11	0	01	01	02	02	03	03	04	04	05	06	07
12	0	01	02	02	03	03	03	04	05	05	06	08
13	0	01	02	02	03	03	04	04	05	06	07	08
14	0	01	02	03	03	04	04	04	05	06	07	09
15	0	01	02	03	03	04	04	05	06	07	08	10
16	01	01	02	03	04	04	05	05	06	07	08	10
17	01	01	02	03	04	04	05	05	07	08	09	11
18	01	01	02	03	04	05	05	05	07	08	09	12
19	01	01	02	04	04	05	05	06	07	08	10	12
20	01	01	03	04	04	05	06	06	08	09	10	13
21	01	01	03	04	05	05	06	07	08	09	11	13
22	01	01	03	04	05	06	06	07	08	10	11	14
23	01	01	03	04	05	06	07	07	09	10	12	15
24	01	02	03	05	05	06	07	08	09	11	12	15
25	01	02	03	05	06	06	07	08	10	11	13	16
26	01	02	03	05	06	07	07	08	10	12	13	17
27	01	02	03	05	06	07	08	09	10	12	14	17
28	01	02	04	05	06	07	08	09	11	13	14	18
29	01	02	04	06	06	07	08	09	11	13	14	18
30	01	02	04	06	07	08	09	10	12	13	15	19
33	01	02	04	06	07	08	09	11	13	15	17	21
63	02	04	08	12	14	16	18	20	24	28	32	40
93	03	06	12	18	21	24	27	30	36	41	48	59

24 Dollars.

	½ per ct.	1 per ct.	2 per ct.	3 per ct.	3½ per ct.	4 per ct.	4½ per ct.	5 per ct.	6 per ct.	7 per ct.	8 per ct.	10 per ct.
Years.												
1	12	24	48	72	84	96	1.08	1.20	1.44	1.68	1.92	2.40
2	24	48	96	1.44	1.68	1.92	2.16	2.40	2.88	3.36	3.84	4.80
3	36	72	1.44	2.16	2.52	2.88	3.24	3.60	4.32	5.04	5.76	7.20
4	48	96	1.92	2.88	3.36	3.84	4.32	4.80	5.76	6.72	7.68	9.60
5	60	1.20	2.40	3.60	4.20	4.80	5.40	6.00	7.20	8.40	9.60	12.00
Months.												
1	01	02	04	06	07	08	09	10	12	14	16	20
2	02	04	08	12	14	16	18	20	24	28	32	40
3	03	06	12	18	21	24	27	30	36	42	48	60
4	04	08	16	24	28	32	36	40	48	56	64	80
5	05	10	20	30	35	40	45	50	60	70	80	1.00
6	06	12	24	36	42	48	54	60	72	84	96	1.20
7	07	14	28	42	49	56	63	70	84	98	1.12	1.40
8	08	16	32	48	56	64	72	80	96	1.12	1.28	1.60
9	09	18	36	54	63	72	81	90	1.08	1.26	1.44	1.80
10	10	20	40	60	70	80	90	1.00	1.20	1.40	1.60	2.00
11	11	22	44	66	77	88	99	1.10	1.32	1.54	1.76	2.20
Days.												
1	0	0	0	0	0	0	0	0	0	0	01	01
2	0	0	0	0	0	01	01	01	01	01	01	01
3	0	0	0	01	01	01	01	01	01	01	02	02
4	0	0	01	01	01	01	01	01	02	02	02	03
5	0	0	01	01	01	01	02	02	02	02	03	03
6	0	0	01	01	01	02	02	02	02	03	03	04
7	0	0	01	01	02	02	02	02	03	03	04	05
8	0	01	01	02	02	02	02	03	03	04	04	05
9	0	01	01	02	02	02	03	03	04	04	05	06
10	0	01	01	02	02	03	03	03	04	05	05	07
11	0	01	01	02	03	03	03	04	04	05	06	07
12	0	01	02	02	03	03	04	04	05	06	06	08
13	0	01	02	03	03	03	04	04	05	06	07	09
14	0	01	02	03	03	04	04	05	06	07	07	09
15	01	01	02	03	04	04	05	05	06	07	08	10
16	01	01	02	03	04	04	05	05	06	07	09	11
17	01	01	02	03	04	05	05	06	07	08	09	11
18	01	01	02	04	04	05	05	06	07	08	10	12
19	01	01	03	04	04	05	06	06	08	09	10	13
20	01	01	03	04	05	05	06	07	08	09	11	13
21	01	01	03	04	05	06	06	07	08	10	11	14
22	01	01	03	04	05	06	07	07	09	10	12	15
23	01	02	03	05	05	06	07	08	09	11	12	15
24	01	02	03	05	06	06	07	08	10	11	13	16
25	01	02	03	05	06	07	08	08	10	12	13	17
26	01	02	03	05	06	07	08	09	10	12	14	17
27	01	02	04	05	06	07	08	09	11	13	14	18
28	01	02	04	06	07	07	08	09	11	13	15	19
29	01	02	04	06	07	08	09	10	12	14	15	19
30	01	02	04	06	07	08	09	10	12	14	16	20
33	01	02	04	07	08	09	10	11	13	15	18	22
63	02	04	08	13	15	17	19	21	25	29	34	42
93	03	06	12	19	22	25	28	31	37	43	50	62

25 Dollars.

	½ per ct.	1 per ct.	2 per ct.	3 per ct.	3½ per ct.	4 per ct.	4½ per ct.	5 per ct.	6 per ct.	7 per ct.	8 per ct.	10 per ct.
Years.												
1	13	25	50	75	88	1.00	1.13	1.25	1.50	1.75	2.00	2.50
2	25	50	1.00	1.50	1.75	2.00	2.25	2.50	3.00	3.50	4.00	5.00
3	38	75	1.50	2.25	2.63	3.00	3.38	3.75	4.50	5.25	6.00	7.50
4	50	1.00	2.00	3.00	3.50	4.00	4.50	5.00	6.00	7.00	8.00	10.00
5	63	1.25	2.50	3.75	4.38	5.00	5.63	6.25	7.50	8.75	10.00	12.50
Months.												
1	01	02	04	06	07	08	09	10	13	15	17	21
2	02	04	08	13	15	17	19	21	25	29	33	42
3	03	06	13	19	22	25	28	31	38	44	50	63
4	04	08	17	25	29	33	38	42	50	58	67	83
5	05	10	21	31	36	42	47	52	63	73	83	1.04
6	06	13	25	38	44	50	56	63	75	88	1.00	1.25
7	07	15	29	44	51	58	66	73	88	1.02	1.17	1.46
8	08	17	33	50	58	67	75	83	1.00	1.17	1.33	1.67
9	09	19	38	56	66	75	84	94	1.13	1.31	1.50	1.88
10	10	21	42	63	73	83	94	1.04	1.25	1.46	1.67	2.08
11	11	23	46	69	80	92	1.03	1.15	1.38	1.60	1.83	2.29
Days.												
1	0	0	0	0	0	0	0	0	0	0	01	01
2	0	0	0	0	0	01	01	01	01	01	01	01
3	0	0	0	01	01	01	01	01	01	01	02	02
4	0	0	01	01	01	01	01	01	02	02	02	03
5	0	0	01	01	01	01	02	02	02	02	03	03
6	0	0	01	01	01	02	02	02	03	03	03	04
7	0	0	01	01	02	02	02	02	03	03	04	05
8	0	01	01	02	02	02	03	03	03	04	04	06
9	0	01	01	02	02	03	03	03	04	04	05	06
10	0	01	01	02	02	03	03	03	04	05	06	07
11	0	01	02	02	03	03	03	04	05	05	06	08
12	0	01	02	03	03	03	04	04	05	06	07	08
13	0	01	02	03	03	04	04	05	05	06	07	09
14	0	01	02	03	03	04	04	05	06	07	08	10
15	01	01	02	03	04	04	05	05	06	07	08	10
16	01	01	02	03	04	04	05	06	07	08	09	11
17	01	01	02	04	04	05	05	06	07	08	09	12
18	01	01	03	04	04	05	06	06	08	09	10	13
19	01	01	03	04	05	05	06	07	08	09	11	13
20	01	01	03	04	05	06	06	07	08	10	11	14
21	01	01	03	04	05	06	07	07	09	10	12	15
22	01	02	03	05	05	06	07	08	09	11	12	15
23	01	02	03	05	06	06	07	08	10	11	13	16
24	01	02	03	05	06	07	08	08	10	12	13	17
25	01	02	03	05	06	07	08	09	10	12	14	17
26	01	02	04	05	06	07	08	09	11	13	14	18
27	01	02	04	06	07	08	08	09	11	13	15	19
28	01	02	04	06	07	08	09	10	12	14	16	19
29	01	02	04	06	07	08	09	10	12	14	16	20
30	01	02	04	06	07	08	09	10	13	15	17	21
33	01	02	05	07	08	09	10	11	14	16	18	23
63	02	04	09	13	15	18	20	22	26	31	35	44
93	03	06	13	19	23	26	29	32	39	45	52	65

26 Dollars.

	½ per ct.	1 per ct.	2 per ct.	3 per ct.	3½ per ct.	4 per ct.	4½ per ct.	5 per ct.	6 per ct.	7 per ct.	8 per ct.	10 per ct.
Years.												
1	13	26	52	78	91	1.04	1.17	1.30	1.56	1.82	2.08	2.60
2	26	52	1.04	1.56	1.82	2.08	2.34	2.60	3.12	3.64	4.16	5.20
3	39	78	1.56	2.34	2.73	3.12	3.51	3.90	4.68	5.46	6.24	7.80
4	52	1.04	2.08	3.12	3.64	4.16	4.68	5.20	6.24	7.28	8.32	10.40
5	65	1.30	2.60	3.90	4.55	5.20	4.85	6.50	7.80	9.10	10.40	13.00
Months.												
1	01	02	04	07	08	09	10	11	13	15	17	22
2	02	04	09	13	15	17	20	22	26	30	35	43
3	03	07	13	20	23	26	29	33	39	46	52	65
4	04	09	17	26	30	35	39	43	52	61	69	87
5	05	11	22	33	38	43	49	54	65	76	87	1.08
6	07	13	26	39	46	52	59	65	78	91	1.04	1.30
7	08	15	30	46	53	61	68	76	91	1.06	1.21	1.52
8	09	17	35	52	61	69	78	87	1.04	1.21	1.39	1.73
9	10	20	39	59	68	78	88	98	1.17	1.37	1.56	1.95
10	11	22	43	65	76	87	98	1.08	1.30	1.52	1.73	2.17
11	12	24	48	72	83	95	1.07	1.19	1.43	1.67	1.91	2.38
Days.												
1	0	0	0	0	0	0	0	0	0	01	01	01
2	0	0	0	0	0	01	01	01	01	01	01	01
3	0	0	0	01	01	01	01	01	01	02	02	02
4	0	0	01	01	01	01	01	01	02	02	03	03
5	0	0	01	01	01	01	02	02	02	03	03	04
6	0	0	01	01	02	02	02	02	03	03	03	04
7	0	01	01	02	02	02	02	03	03	04	04	05
8	0	01	01	02	02	02	03	03	03	04	05	06
9	0	01	01	02	02	03	03	03	04	05	05	07
10	0	01	01	02	03	03	03	04	04	05	06	07
11	0	01	02	02	03	03	04	04	05	06	06	08
12	0	01	02	03	03	03	04	04	05	06	07	09
13	0	01	02	03	03	04	04	05	06	07	08	09
14	01	01	02	03	04	04	05	05	06	07	08	10
15	01	01	02	03	04	04	05	05	07	08	09	11
16	01	01	02	03	04	05	05	06	07	08	09	12
17	01	01	02	04	04	05	06	06	07	09	10	12
18	01	01	03	04	05	05	06	07	08	09	10	13
19	01	01	03	04	05	05	06	07	08	10	11	14
20	01	01	03	04	05	06	07	07	09	10	12	14
21	01	02	03	05	05	06	07	08	09	11	12	15
22	01	02	03	05	06	06	07	08	10	11	13	16
23	01	02	03	05	06	07	07	08	10	12	13	17
24	01	02	03	05	06	07	08	09	10	12	14	17
25	01	02	04	05	06	07	08	09	11	13	14	18
26	01	02	04	06	07	08	08	09	11	13	15	19
27	01	02	04	06	07	08	09	10	12	14	16	20
28	01	02	04	06	07	08	09	10	12	14	16	20
29	01	02	04	06	07	08	09	10	13	15	17	21
30	01	02	04	07	08	09	10	11	13	15	17	22
33	01	02	05	07	08	10	12	12	14	17	19	24
63	02	05	09	14	16	18	20	23	27	32	36	45
93	03	07	13	20	24	27	30	34	40	47	54	67

27 Dollars.

Years.	½ per ct.	1 per ct.	2 per ct.	3 per ct.	3½ per ct.	4 per ct.	4½ per ct.	5 per ct.	6 per ct.	7 per ct.	8 per ct.	10 per ct.
1	14	27	54	81	95	1.08	1.22	1.35	1.62	1.89	2.16	2.70
2	27	54	1.08	1.62	1.89	2.16	2.43	2.70	3.24	3.78	4.32	5.40
3	41	81	1.62	2.43	2.84	3.24	3.65	4.05	4.86	5.67	6.48	8.10
4	54	1.08	2.16	3.24	3.78	4.32	4.86	5.40	6.48	7.56	8.64	10.80
5	68	1.35	2.70	4.05	4.73	5.40	6.08	6.75	8.10	9.45	10.80	13.50
Months.												
1	01	02	05	07	08	09	10	11	14	16	18	23
2	02	05	09	14	16	18	20	23	27	32	36	45
3	03	07	14	20	24	27	30	34	41	47	54	68
4	05	09	18	27	32	36	41	45	54	63	72	90
5	06	11	23	34	39	45	51	56	68	79	90	1.13
6	07	14	27	41	47	54	61	68	81	95	1.08	1.35
7	08	16	32	47	55	63	71	79	95	1.10	1.26	1.58
8	09	18	36	54	63	72	81	90	1.08	1.26	1.44	1.80
9	10	20	41	61	71	81	91	1.01	1.22	1.42	1.62	2.03
10	11	23	45	68	79	90	1.01	1.13	1.35	1.58	1.80	2.25
11	12	25	50	74	87	99	1.11	1.24	1.49	1.73	1.98	2.48
Days.												
1	0	0	0	0	0	0	0	0	0	01	01	01
2	0	0	0	0	01	01	01	01	01	01	01	01
3	0	0	0	01	01	01	01	01	01	02	02	02
4	0	0	01	01	01	01	01	02	02	02	02	03
5	0	0	01	01	01	02	02	02	02	03	03	04
6	0	0	01	01	02	02	02	02	03	03	04	05
7	0	01	01	02	02	02	02	03	03	04	04	05
8	0	01	01	02	02	02	03	03	04	04	05	06
9	0	01	01	02	02	03	03	03	04	05	05	07
10	0	01	02	02	03	03	03	04	05	05	06	08
11	0	01	02	02	03	03	04	04	05	06	07	08
12	0	01	02	03	03	04	04	05	05	06	07	09
13	0	01	02	03	03	04	04	05	06	07	08	10
14	0	01	02	03	04	04	05	05	06	07	08	11
15	01	01	02	03	04	05	05	06	07	08	09	11
16	01	01	02	04	04	05	05	06	07	08	10	12
17	01	01	03	04	04	05	06	06	08	09	10	13
18	01	01	03	04	05	05	06	07	08	09	11	14
19	01	01	03	04	05	06	06	07	09	10	11	14
20	01	02	03	05	05	06	07	08	09	11	12	15
21	01	02	03	05	06	06	07	08	09	11	13	16
22	01	02	03	05	06	07	07	08	10	12	13	17
23	01	02	03	05	06	07	08	09	10	12	14	17
24	01	02	04	05	06	07	08	09	11	13	14	18
25	01	02	04	06	07	08	08	09	11	13	15	19
26	01	02	04	06	07	08	09	10	12	14	16	20
27	01	02	04	06	07	08	09	10	12	14	16	20
28	01	02	04	06	07	08	09	11	13	15	17	21
29	01	02	04	07	08	09	10	11	13	15	17	22
30	01	02	05	07	08	09	10	11	14	16	18	23
33	01	02	05	07	09	10	11	12	15	17	20	25
63	02	05	09	14	17	19	21	24	28	33	38	47
93	03	07	14	21	24	28	31	35	42	49	56	70

28 Dollars.

Years.	½ per ct.	1 per ct.	2 per ct.	3 per ct.	3½ per ct.	4 per ct.	4½ per ct.	5 per ct.	6 per ct.	7 per ct.	8 per ct.	10 per ct.
1	14	28	56	84	98	1.12	1.26	1.40	1.68	1.96	2.24	2.80
2	28	56	1.12	1.68	1.96	2.24	2.52	2.80	3.36	3.92	4.48	5.60
3	42	84	1.68	2.52	2.94	3.36	3.78	4.20	5.04	5.88	6.72	8.40
4	56	1.12	2.24	3.36	3.92	4.48	5.04	5.60	6.72	7.84	8.96	11.20
5	70	1.40	2.80	4.20	4.90	5.60	6.30	7.00	8.40	9.80	11.20	14.00
Months.												
1	01	02	05	07	08	09	11	12	14	16	19	23
2	02	05	09	14	16	19	21	23	28	33	37	47
3	04	07	14	21	25	28	32	35	42	49	56	70
4	05	09	19	28	33	37	42	47	56	65	75	93
5	06	12	23	35	41	47	53	58	70	82	93	1.17
6	07	14	28	42	49	56	63	70	84	98	1.12	1.40
7	08	16	33	49	57	65	74	82	98	1.14	1.31	1.63
8	09	19	37	56	65	75	84	93	1.12	1.31	1.49	1.87
9	11	21	42	63	74	84	95	1.05	1.26	1.47	1.68	2.10
10	12	23	47	70	82	93	1.05	1.17	1.40	1.63	1.87	2.33
11	13	26	51	77	90	1.03	1.16	1.28	1.54	1.80	2.05	2.57
Days.												
1	0	0	0	0	0	0	0	0	0	01	01	01
2	0	0	0	0	01	01	01	01	01	01	01	02
3	0	0	0	01	01	01	01	01	01	02	02	02
4	0	0	01	01	01	01	01	02	02	02	02	03
5	0	0	01	01	01	02	02	02	02	03	03	04
6	0	0	01	01	02	02	02	02	03	03	04	05
7	0	01	01	02	02	02	02	03	03	04	04	05
8	0	01	01	02	02	02	03	03	04	04	05	06
9	0	01	01	02	02	03	03	04	04	05	06	07
10	0	01	02	02	03	03	04	04	05	05	06	08
11	0	01	02	03	03	03	04	04	05	06	07	09
12	0	01	02	03	03	04	04	05	06	07	07	09
13	01	01	02	03	04	04	05	05	06	07	08	10
14	01	01	02	03	04	04	05	05	07	08	09	11
15	01	01	02	04	04	05	05	06	07	08	09	12
16	01	01	02	04	04	05	06	06	07	09	10	12
17	01	01	03	04	05	05	06	07	08	09	11	13
18	01	01	03	04	05	06	06	07	08	10	11	14
19	01	01	03	04	05	06	07	07	09	10	12	15
20	01	02	03	05	05	06	07	08	09	11	12	16
21	01	02	03	05	06	07	07	08	10	11	13	16
22	01	02	03	05	06	07	08	09	10	12	14	17
23	01	02	04	05	06	07	08	09	11	13	14	18
24	01	02	04	06	07	07	08	09	11	13	15	19
25	01	02	04	06	07	08	09	10	12	14	16	19
26	01	02	04	06	07	08	09	10	12	14	16	20
27	01	02	04	06	07	08	09	11	13	15	17	21
28	01	02	04	07	08	09	10	11	13	15	17	22
29	01	02	05	07	08	09	10	11	14	16	18	23
30	01	02	05	07	08	09	11	12	14	16	19	23
33	01	03	05	08	09	10	12	13	15	18	21	26
63	02	05	10	15	17	20	22	25	29	34	39	49
93	04	07	14	22	25	29	83	36	43	51	58	72

29 Dollars.

Years.	½ per ct.	1 per ct.	2 per ct.	3 per ct.	3½ per ct.	4 per ct.	4½ per ct.	5 per ct.	6 per ct.	7 per ct.	8 per ct.	10 per ct.
1	15	29	58	87	1.02	1.16	1.31	1.45	1.74	2.03	2.32	2.90
2	29	58	1.16	1.74	2.03	2.32	2.61	2.90	3.48	4.06	4.64	5.80
3	44	87	1.74	2.61	3.05	3.48	3.92	4.35	5.22	6.09	6.96	8.70
4	58	1.16	2.32	3.48	4.06	4.64	5.22	5.80	6.96	8.12	9.28	11.60
5	73	1.45	2.90	4.35	5.08	5.80	6.53	7.25	8.70	10.15	11.60	14.50
Months.												
1	01	02	05	07	08	10	11	12	15	17	19	24
2	02	05	10	15	17	19	22	24	29	34	39	48
3	04	07	15	22	25	29	33	36	44	51	58	73
4	05	10	19	29	34	39	44	48	58	68	77	97
5	06	12	24	36	42	48	54	60	73	85	97	1.21
6	07	15	29	44	51	58	65	73	87	1.02	1.16	1.45
7	08	17	34	51	59	68	76	85	1.02	1.18	1.35	1.69
8	10	19	39	58	68	77	87	97	1.16	1.35	1.55	1.93
9	11	22	44	65	76	87	98	1.09	1.31	1.52	1.74	2.18
10	12	24	48	73	85	97	1.09	1.21	1.45	1.69	1.93	2.42
11	13	27	53	80	93	1.06	1.20	1.33	1.60	1.86	2.13	2.66
Days.												
1	0	0	0	0	0	0	0	0	0	01	01	01
2	0	0	0	0	01	01	01	01	01	01	01	02
3	0	0	0	01	01	01	01	01	01	02	02	02
4	0	0	01	01	01	01	01	02	02	02	03	03
5	0	0	01	01	01	02	02	02	02	03	03	04
6	0	0	01	01	02	02	02	02	03	03	04	05
7	0	01	01	02	02	02	03	03	03	04	05	06
8	0	01	01	02	02	03	03	03	04	05	05	06
9	0	01	01	02	03	03	03	04	04	05	06	07
10	0	01	02	02	03	03	04	04	05	06	06	08
11	0	01	02	03	03	04	04	04	05	06	07	09
12	0	01	02	03	03	04	04	05	06	07	08	10
13	01	01	02	03	04	04	05	05	06	07	08	10
14	01	01	02	03	04	05	05	06	07	08	09	11
15	01	01	02	04	04	05	05	06	07	08	10	12
16	01	01	03	04	05	05	06	06	08	09	10	13
17	01	01	03	04	05	05	06	07	08	10	11	14
18	01	01	03	04	05	06	07	07	09	10	12	15
19	01	02	03	05	05	06	07	08	09	11	12	15
20	01	02	03	05	06	06	07	08	10	11	13	16
21	01	02	03	05	06	07	08	08	10	12	14	17
22	01	02	04	05	06	07	08	09	11	12	14	18
23	01	02	04	06	06	07	08	09	11	13	15	19
24	01	02	04	06	07	08	09	10	12	14	15	19
25	01	02	04	06	07	08	09	10	12	14	16	20
26	01	02	04	06	07	08	09	10	13	15	17	21
27	01	02	04	07	08	09	10	11	13	15	17	22
28	01	02	05	07	08	09	10	11	14	16	18	23
29	01	02	05	07	08	09	11	12	14	16	19	23
30	01	02	05	07	08	10	11	12	15	17	19	24
33	01	03	05	08	09	11	12	13	16	19	21	27
63	03	05	10	15	18	20	23	25	30	35	40	50
93	04	07	15	22	26	30	34	37	45	53	60	75

30 Dollars.

	½ per ct.	1 per ct.	2 per ct.	3 per ct.	3½ per ct.	4 per ct.	4½ per ct.	5 per ct.	6 per ct.	7 per ct.	8 per ct.	10 per ct.
Years.												
1	15	30	60	90	1.05	1.20	1.35	1.50	1.80	2.10	2.40	3.00
2	30	60	1.20	1.80	2.10	2.40	2.70	3.00	3.60	4.20	4.80	6.00
3	45	90	1.80	2.70	3.15	3.60	4.05	4.50	5.40	6.30	7.20	9.00
4	60	1.20	2.40	3.60	4.20	4.80	5.40	6.00	7.20	8.40	9.60	12.00
5	75	1.50	3.00	4.50	5.25	6.00	6.75	7.50	9.00	10.50	12.00	15.00
Months.												
1	01	03	05	08	09	10	11	13	15	18	20	25
2	03	05	10	15	18	20	23	25	30	35	40	50
3	04	08	15	23	26	30	34	38	45	53	60	75
4	05	10	20	30	35	40	45	50	60	70	80	1.00
5	06	13	25	38	44	50	56	63	75	88	1.00	1.25
6	08	15	30	45	53	60	68	75	90	1.05	1.20	1.50
7	09	18	35	53	61	70	79	88	1.05	1.23	1.40	1.75
8	10	20	40	60	70	80	90	1.00	1.20	1.40	1.60	2.00
9	11	23	45	68	79	90	1.01	1.13	1.35	1.58	1.80	2.25
10	13	25	50	75	88	1.00	1.13	1.25	1.50	1.75	2.00	2.50
11	14	28	55	83	96	1.10	1.24	1.38	1.65	1.93	2.20	2.75
Days.												
1	0	0	0	0	0	0	0	0	01	01	01	01
2	0	0	0	01	01	01	01	01	01	01	01	02
3	0	0	01	01	01	01	01	01	02	02	02	03
4	0	0	01	01	01	01	02	02	02	02	03	03
5	0	0	01	01	01	02	02	02	03	03	03	04
6	0	01	01	02	02	02	02	03	03	04	04	05
7	0	01	01	02	02	02	03	03	04	04	05	06
8	0	01	01	02	02	03	03	03	04	05	05	07
9	0	01	02	02	03	03	03	04	05	05	06	08
10	0	01	02	03	03	03	04	04	05	06	07	08
11	0	01	02	03	03	04	04	05	06	06	07	09
12	01	01	02	03	04	04	05	05	06	07	08	10
13	01	01	02	03	04	04	05	05	07	08	09	11
14	01	01	02	04	04	05	05	06	07	08	09	12
15	01	01	03	04	04	05	06	06	08	09	10	13
16	01	01	03	04	05	05	06	07	08	09	11	13
17	01	01	03	04	05	06	06	07	09	10	11	14
18	01	02	03	05	05	06	07	08	09	11	12	15
19	01	02	03	05	06	06	07	08	10	11	13	16
20	01	02	03	05	06	07	08	08	10	12	13	17
21	01	02	04	05	06	07	08	09	11	12	14	18
22	01	02	04	06	06	07	08	09	11	13	15	18
23	01	02	04	06	07	08	09	10	12	13	15	19
24	01	02	04	06	07	08	09	10	12	14	16	20
25	01	02	04	06	07	08	09	10	13	15	17	21
26	01	02	04	07	08	09	10	11	13	15	17	22
27	01	02	05	07	08	09	10	11	14	16	18	23
28	01	02	05	07	08	09	11	12	14	16	19	23
29	01	02	05	07	08	10	11	12	15	17	19	24
30	01	03	05	08	09	10	11	13	15	18	20	25
33	01	03	06	08	10	11	12	14	17	20	22	28
63	03	05	11	16	18	21	24	26	32	37	42	53
93	04	08	16	23	27	31	35	39	47	55	62	78

31 Dollars.

Years.	½ per ct.	1 per ct.	2 per ct.	3 per ct.	3½ per ct.	4 per ct.	4½ per ct.	5 per ct.	6 per ct.	7 per et.	8 per ct.	10 per ct.
1	16	31	62	93	1.09	1.24	1.40	1.55	1.86	2.17	2.48	3.10
2	31	62	1.24	1.86	2.17	2.48	2.79	3.10	3.72	4.34	4.96	6.20
3	47	93	1.86	2.79	3.26	3.72	4.19	4.65	5.58	6.51	7.44	9.30
4	62	1.24	2.48	3.72	4.34	4.96	5.58	6.20	7.44	8.68	9.92	12.40
5	78	1.55	3.10	4.65	5.43	6.20	6.98	7.75	9.30	10.85	12.40	15.50
Months.												
1	01	03	05	08	09	10	12	13	16	18	21	26
2	03	05	10	16	18	21	23	26	31	36	41	52
3	04	08	16	23	27	31	35	39	47	54	62	78
4	05	10	21	31	36	41	47	52	62	72	83	1.03
5	06	13	26	39	45	52	58	65	78	90	1.03	1.29
6	08	16	31	47	54	62	70	78	93	1.09	1.24	1.55
7	09	18	36	54	63	72	81	90	1.09	1.27	1.45	1.81
8	10	21	41	62	72	83	93	1.03	1.24	1.45	1.65	2.07
9	12	23	47	70	81	93	1.05	1.16	1.40	1.63	1.86	2.33
10	13	26	52	78	90	1.03	1.16	1.29	1.55	1.81	2.07	2.58
11	14	28	57	85	99	1.14	1.28	1.42	1.71	1.99	2.27	2.84
Days.												
1	0	0	0	0	0	0	0	0	01	01	01	01
2	0	0	0	01	01	01	01	01	01	01	01	02
3	0	0	01	01	01	01	01	01	02	02	02	03
4	0	0	01	01	01	01	02	02	02	02	03	03
5	0	0	01	01	02	02	02	02	03	03	03	04
6	0	01	01	02	02	02	02	03	03	04	04	05
7	0	01	01	02	02	02	03	03	04	04	05	06
8	0	01	01	02	02	03	03	03	04	05	06	07
9	0	01	02	02	03	03	03	04	05	05	06	08
10	0	01	02	03	03	03	04	04	05	06	07	09
11	0	01	02	03	03	04	04	05	06	07	08	09
12	01	01	02	03	04	04	05	05	06	07	08	10
13	01	01	02	03	04	04	05	06	07	08	09	11
14	01	01	02	04	04	05	05	06	07	08	10	12
15	01	01	03	04	05	05	06	06	08	09	10	13
16	01	01	03	04	05	06	06	07	08	09	11	14
17	01	01	03	04	05	06	07	07	09	10	12	15
18	01	02	03	05	05	06	07	08	09	11	12	16
19	01	02	03	05	06	07	07	08	10	11	13	16
20	01	02	03	05	06	07	08	09	10	12	14	17
21	01	02	04	05	06	07	08	09	11	13	14	18
22	01	02	04	06	07	08	09	09	11	13	15	19
23	01	02	04	06	07	08	09	10	12	14	16	20
24	01	02	04	06	07	08	09	10	12	14	17	21
25	01	02	04	06	08	09	10	11	13	15	17	22
26	01	02	04	07	08	09	10	11	13	16	18	22
27	01	02	05	07	08	09	10	12	14	16	19	23
28	01	02	05	07	08	10	11	12	14	17	19	24
29	01	02	05	07	09	10	11	12	15	17	20	25
30	01	03	05	08	09	10	12	13	16	18	21	26
33	01	03	06	09	10	11	13	14	17	20	23	28
63	03	05	11	16	19	22	24	27	33	38	43	54
93	04	08	16	24	28	32	36	40	48	56	64	80

32 Dollars.

	½ per ct.	1 per ct.	2 per ct.	3 per ct.	3½ per ct.	4 per ct.	4½ per ct.	5 per ct.	6 per ct.	7 per ct.	8 per ct.	10 per ct.
Years.												
1	16	32	64	96	1.12	1.28	1.44	1.60	1.92	2.24	2.56	3.20
2	32	64	1.28	1.92	2.24	2.56	2.88	3.20	3.84	4.48	5.12	6.40
3	48	96	1.92	2.88	3.36	3.84	4.32	4.80	5.76	6.72	7.68	9.60
4	64	1.28	2.56	3.84	4.48	5.12	5.76	6.40	7.68	8.96	10.24	12.80
5	80	1.60	3.20	4.80	5.60	6.40	7.20	8.00	9.60	11.20	12.80	16.00
Months.												
1	01	03	05	08	09	11	12	13	16	19	21	27
2	03	05	11	16	19	21	24	27	32	37	43	53
3	04	08	16	24	28	32	36	40	48	56	64	80
4	05	11	21	32	37	43	48	53	64	75	85	1.07
5	07	13	27	40	47	53	60	67	80	93	1.07	1.33
6	08	16	32	48	56	64	72	80	96	1.12	1.28	1.60
7	09	19	37	56	65	75	84	93	1.12	1.31	1.49	1.87
8	11	21	43	64	75	85	96	1.07	1.28	1.49	1.71	2.13
9	12	24	48	72	84	96	1.08	1.20	1.44	1.68	1.92	2.40
10	13	27	53	80	93	1.07	1.20	1.33	1.60	1.87	2.13	2.67
11	15	29	59	88	1.03	1.17	1.32	1.47	1.76	2.05	2.35	2.93
Days.												
1	0	0	0	0	0	0	0	0	01	01	01	01
2	0	0	0	01	01	01	01	01	01	01	01	02
3	0	0	01	01	01	01	01	01	02	02	02	03
4	0	0	01	01	01	01	02	02	02	02	03	04
5	0	0	01	01	02	02	02	02	03	03	04	04
6	0	01	01	02	02	02	02	03	03	04	04	05
7	0	01	01	02	02	02	03	03	04	04	05	06
8	0	01	01	02	02	03	03	04	04	05	06	07
9	0	01	02	02	03	03	04	04	05	06	07	08
10	0	01	02	03	03	04	04	04	05	06	07	09
11	0	01	02	03	03	04	04	05	06	07	08	10
12	01	01	02	03	04	04	05	05	06	07	09	11
13	01	01	02	03	04	05	05	06	07	08	09	12
14	01	01	02	04	04	05	06	06	07	09	10	12
15	01	01	03	04	05	05	06	07	08	09	11	13
16	01	01	03	04	05	06	06	07	09	10	11	14
17	01	02	03	05	05	06	07	08	09	11	12	15
18	01	02	03	05	06	06	07	08	10	11	13	16
19	01	02	03	05	06	07	08	08	10	12	14	17
20	01	02	04	05	06	07	08	09	11	12	14	18
21	01	02	04	06	07	07	08	09	11	13	15	19
22	01	02	04	06	07	08	09	10	12	14	16	20
23	01	02	04	06	07	08	09	10	12	14	16	20
24	01	02	04	06	07	09	10	11	13	15	17	21
25	01	02	04	07	08	09	10	11	13	16	18	22
26	01	02	05	07	08	09	10	12	14	16	18	23
27	01	02	05	07	08	10	11	12	14	17	19	24
28	01	02	05	07	09	10	11	12	15	17	20	25
29	01	03	05	08	09	10	12	13	15	18	21	26
30	01	03	05	08	09	11	12	13	16	19	21	27
33	01	03	06	09	10	12	13	15	18	21	23	29
63	03	06	11	17	20	22	25	28	34	39	45	56
93	04	08	17	25	29	33	37	41	50	58	66	83

33 Dollars.

Years.	½ per ct.	1 per ct.	2 per ct.	3 per ct.	3½ per ct.	4 per ct.	4½ per ct.	5 per ct.	6 per ct.	7 per ct.	8 per ct.	10 per ct.
1	17	33	66	99	1.16	1.32	1.49	1.65	1.98	2.31	2.64	3.30
2	33	66	1.32	1.98	2.31	2.64	2.97	3.30	3.96	4.62	5.28	6.60
3	50	99	1.98	2.97	3.47	3.96	4.46	4.95	5.94	6.93	7.92	9.90
4	66	1.32	2.64	3.96	4.62	5.28	5.94	6.60	7.92	9.24	10.56	13.20
5	83	1.65	3.30	4.95	5.78	6.60	7.43	8.25	9.90	11.55	13.20	16.50
Months.												
1	01	03	06	08	10	11	12	14	17	19	22	28
2	03	06	11	17	19	22	25	28	33	39	44	55
3	04	08	17	25	29	33	37	41	50	58	66	83
4	06	11	22	33	39	44	50	55	66	77	88	1.10
5	07	14	28	41	48	55	62	69	83	96	1.10	1.38
6	08	17	33	50	58	66	74	83	99	1.16	1.32	1.65
7	10	19	39	58	67	77	87	96	1.16	1.35	1.54	1.93
8	11	22	44	66	77	88	99	1.10	1.32	1.54	1.76	2.20
9	12	25	50	74	87	99	1.11	1.24	1.49	1.73	1.98	2.48
10	14	28	55	83	96	1.10	1.24	1.38	1.65	1.93	2.20	2.75
11	15	30	61	91	1.06	1.21	1.36	1.51	1.82	2.12	2.42	3.03
Days.												
1	0	0	0	0	0	0	0	0	01	01	01	01
2	0	0	0	01	01	01	01	01	01	01	01	02
3	0	0	01	01	01	01	01	01	02	02	02	03
4	0	0	01	01	01	01	02	02	02	03	03	04
5	0	0	01	01	02	02	02	02	03	03	04	05
6	0	01	01	02	02	02	02	03	03	04	04	06
7	0	01	01	02	02	03	03	03	04	04	05	06
8	0	01	01	02	03	03	03	04	04	05	06	07
9	0	01	02	02	03	03	04	04	05	06	07	08
10	0	01	02	03	03	04	04	05	06	06	07	09
11	01	01	02	03	04	04	05	05	06	07	08	10
12	01	01	02	03	04	04	05	06	07	08	09	11
13	01	01	02	04	04	05	05	06	07	08	10	12
14	01	01	03	04	04	05	06	06	08	09	10	13
15	01	01	03	04	05	06	06	07	08	10	11	14
16	01	01	03	04	05	06	07	07	09	10	12	15
17	01	02	03	05	05	06	07	08	09	11	12	16
18	01	02	03	05	06	07	07	08	10	12	13	17
19	01	02	03	05	06	07	08	09	10	12	14	17
20	01	02	04	06	06	07	08	09	11	13	15	18
21	01	02	04	06	07	08	09	10	12	13	15	19
22	01	02	04	06	07	08	09	10	12	14	16	20
23	01	02	04	06	07	08	09	11	13	15	17	21
24	01	02	04	07	08	09	10	11	13	15	18	22
25	01	02	05	07	08	09	10	11	14	16	18	23
26	01	02	05	07	08	10	11	12	14	17	19	24
27	01	02	05	07	09	10	11	12	15	17	20	25
28	01	03	05	08	09	10	12	13	15	18	21	26
29	01	03	05	08	09	11	12	13	16	19	21	27
30	01	03	06	08	10	11	12	14	17	19	22	28
33	02	03	06	09	11	12	14	15	18	21	24	30
63	03	06	12	17	20	23	26	29	35	40	46	58
93	04	09	17	26	30	34	38	43	51	60	68	85

34 Dollars.

	½ per ct.	1 per ct.	2 per ct.	3 per ct.	3½ per ct.	4 per ct.	4½ per ct.	5 per ct.	6 per ct.	7 per ct.	8 per ct.	10 per ct.
Years.												
1	17	34	68	1.02	1.19	1.36	1.53	1.70	2.04	2.38	2.72	3.40
2	34	68	1.36	2.04	2.38	2.72	3.06	3.40	4.08	4.76	5.44	6.80
3	51	1.02	2.04	3.06	3.57	4.08	4.59	5.10	6.12	7.14	8.16	10.20
4	68	1.36	2.72	4.08	4.76	5.44	6.12	6.80	8.16	9.52	10.88	13.60
5	85	1.70	3.40	5.10	5.95	6.80	7.65	8.50	10.20	11.90	13.60	17.00
Months.												
1	01	03	06	09	10	11	13	14	17	20	23	28
2	03	06	11	17	20	23	26	28	34	40	45	57
3	04	09	17	26	30	34	38	43	51	60	68	85
4	06	11	23	34	40	45	51	57	68	79	91	1.13
5	07	14	28	43	50	57	64	71	85	99	1.13	1.42
6	09	17	34	51	60	68	77	85	1.02	1.19	1.36	1.70
7	10	20	40	60	69	79	89	99	1.19	1.39	1.59	1.98
8	11	23	45	68	79	91	1.02	1.13	1.36	1.59	1.81	2.27
9	13	26	51	77	89	1.02	1.15	1.28	1.53	1.79	2.04	2.55
10	14	28	57	85	99	1.13	1.28	1.42	1.70	1.98	2.27	2.83
11	16	31	62	94	1.09	1.25	1.40	1.56	1.87	2.18	2.49	3.12
Days.												
1	0	0	0	0	0	0	0	0	01	01	01	01
2	0	0	0	01	01	01	01	01	01	01	02	02
3	0	0	01	01	01	01	01	01	02	02	02	03
4	0	0	01	01	01	02	02	02	02	03	03	04
5	0	0	01	01	02	02	02	02	03	03	04	05
6	0	01	01	02	02	02	03	03	03	04	05	06
7	0	01	01	02	02	03	03	03	04	05	05	07
8	0	01	02	02	03	03	03	04	05	05	06	08
9	0	01	02	03	03	03	04	04	05	06	07	09
10	0	01	02	03	03	04	04	05	06	07	08	09
11	01	01	02	03	04	04	05	05	06	07	08	10
12	01	01	02	03	04	05	05	06	07	08	09	11
13	01	01	02	04	04	05	06	06	07	09	10	12
14	01	01	03	04	05	05	06	07	08	09	11	13
15	01	01	03	04	05	06	06	07	09	10	11	14
16	01	02	03	05	05	06	07	08	09	11	12	15
17	01	02	03	05	06	06	07	08	10	11	13	16
18	01	02	03	05	06	07	08	09	10	12	14	17
19	01	02	04	05	06	07	08	09	11	13	14	18
20	01	02	04	06	07	08	09	09	11	13	15	19
21	01	02	04	06	07	08	09	10	12	14	16	20
22	01	02	04	06	07	08	09	10	12	15	17	21
23	01	02	04	07	08	09	10	11	13	15	17	22
24	01	02	05	07	08	09	10	11	14	16	18	23
25	01	02	05	07	08	09	11	12	14	17	19	24
26	01	02	05	07	09	10	11	12	15	17	20	25
27	01	03	05	08	09	10	11	13	15	18	20	26
28	01	03	05	08	09	11	12	13	16	19	21	26
29	01	03	05	08	10	11	12	14	16	19	22	27
30	01	03	06	09	10	11	13	14	17	20	23	28
33	02	03	06	09	11	12	14	16	19	22	25	31
63	03	06	12	18	21	24	27	30	36	42	48	60
93	04	09	18	26	31	35	40	44	53	61	70	88

43

35 Dollars.

Years.	½ per ct.	1 per ct.	2 per ct.	3 per ct.	3½ per ct.	4 per ct.	4½ per ct.	5 per ct.	6 per ct.	7 per ct.	8 per ct.	10 per ct.
1	18	35	70	1.05	1.23	1.40	1.58	1.75	2.10	2.45	2.80	3.50
2	35	70	1.40	2.10	2.45	2.80	3.15	3.50	4.20	4.90	5.60	7.00
3	53	1.05	2.10	3.15	3.68	4.20	5.73	5.25	6.30	7.35	8.40	10.50
4	70	1.40	2.80	4.20	4.90	5.60	6.30	7.00	8.40	9.80	11.20	14.00
5	88	1.75	3.50	5.25	6.13	7.00	7.88	8.75	10.50	12.25	14.00	17.50
Months.												
1	01	03	06	09	10	12	13	15	18	20	23	29
2	03	06	12	18	20	23	26	29	35	41	47	58
3	04	09	18	26	31	35	39	44	53	61	70	88
4	06	12	23	35	41	47	53	58	70	82	93	1.17
5	07	15	29	44	51	58	66	73	88	1.02	1.17	1.46
6	09	18	35	53	61	70	79	88	1.05	1.23	1.40	1.75
7	10	20	41	61	71	82	92	1.02	1.23	1.43	1.63	2.04
8	12	23	47	70	82	93	1.05	1.17	1.40	1.63	1.87	2.33
9	13	26	53	79	92	1.05	1.18	1.31	1.58	1.84	2.10	2.63
10	15	29	58	88	1.02	1.17	1.31	1.46	1.75	2.04	2.33	2.92
11	16	32	64	96	1.12	1.28	1.44	1.60	1.93	2.25	2.57	3.21
Days.												
1	0	0	0	0	0	0	0	0	01	01	01	01
2	0	0	0	01	01	01	01	01	01	01	02	02
3	0	0	01	01	01	01	01	01	02	02	02	03
4	0	0	01	01	01	02	02	02	02	03	03	04
5	0	0	01	01	02	02	02	02	03	03	04	05
6	0	01	01	02	02	02	03	03	04	04	05	06
7	0	01	01	02	02	03	03	03	04	05	05	07
8	0	01	02	02	03	03	04	04	05	05	06	08
9	0	01	02	03	03	04	04	04	05	06	07	09
10	0	01	02	03	03	04	04	05	06	07	08	10
11	01	01	02	03	04	04	05	05	06	07	09	11
12	01	01	02	04	04	05	05	06	07	08	09	12
13	01	01	03	04	04	05	06	06	08	09	10	13
14	01	01	03	04	05	05	06	07	08	10	11	14
15	01	01	03	04	05	06	07	07	09	10	12	15
16	01	02	03	05	05	06	07	08	09	11	12	16
17	01	02	03	05	06	07	07	08	10	12	13	17
18	01	02	04	05	06	07	08	09	11	12	14	18
19	01	02	04	06	06	07	08	09	11	13	15	18
20	01	02	04	06	07	08	09	10	12	14	16	19
21	01	02	04	06	07	08	09	10	12	14	16	20
22	01	02	04	06	07	09	10	11	13	15	17	21
23	01	02	04	07	08	09	10	11	13	16	18	22
24	01	02	05	07	08	09	11	12	14	16	19	23
25	01	02	05	07	09	10	11	12	15	17	19	24
26	01	03	05	08	09	10	11	13	15	18	20	25
27	01	03	05	08	09	11	12	13	16	18	21	26
28	01	03	05	08	10	11	12	14	16	19	22	27
29	01	03	06	08	10	11	13	14	17	20	23	28
30	01	03	06	09	10	12	13	15	18	20	23	29
33	02	03	06	10	11	13	14	16	19	22	26	32
63	03	06	12	18	21	25	28	31	37	43	49	61
93	05	09	18	27	32	36	41	45	54	63	72	90

44

36 Dollars.

	½ per ct.	1 per ct.	2 per ct.	3 per ct.	3½ per ct.	4 per ct.	4½ per ct.	5 per ct.	6 per ct.	7 per ct.	8 per ct.	10 per ct.
Years.												
1	18	36	72	1.08	1.26	1.44	1.62	1.80	2.16	2.52	2.88	3.60
2	36	72	1.44	2.16	2.52	2.88	3.24	3.60	4.32	5.04	5.76	7.20
3	54	1.08	2.16	3.24	3.78	4.32	4.86	5.40	6.48	7.56	8.64	10.80
4	72	1.44	2.88	4.32	5.04	5.76	6.48	7.20	8.64	10.08	11.52	14.40
5	90	1.80	3.60	5.40	6.30	7.20	8.10	9.00	10.80	12.60	14.40	18.00
Months.												
1	02	03	06	09	11	12	14	15	18	21	24	30
2	03	06	12	18	21	24	27	30	36	42	48	60
3	05	09	18	27	32	36	41	45	54	63	72	90
4	06	12	24	36	42	48	54	60	72	84	96	1.20
5	08	15	30	45	53	60	68	75	90	1.05	1.20	1.50
6	09	18	36	54	63	72	81	90	1.08	1.26	1.44	1.80
7	11	21	42	63	74	84	95	1.05	1.26	1.47	1.68	2.10
8	12	24	48	72	84	96	1.08	1.20	1.44	1.68	1.92	2.40
9	14	27	54	81	95	1.08	1.22	1.35	1.62	1.89	2.16	2.70
10	15	30	60	90	1.05	1.20	1.35	1.50	1.80	2.10	2.40	3.00
11	17	33	66	99	1.16	1.32	1.49	1.65	1.98	2.31	2.64	3.30
Days.												
1	0	0	0	0	0	0	0	01	01	01	01	01
2	0	0	0	01	01	01	01	01	01	01	02	02
3	0	0	01	01	01	01	01	02	02	02	02	03
4	0	0	01	01	01	02	02	02	02	03	03	04
5	0	01	01	02	02	02	02	03	03	04	04	05
6	0	01	01	02	02	02	03	03	04	04	05	06
7	0	01	01	02	02	03	03	04	04	05	06	07
8	0	01	02	02	03	03	04	04	05	06	06	08
9	0	01	02	03	03	04	04	05	05	06	07	09
10	01	01	02	03	04	04	05	05	06	07	08	10
11	01	01	02	03	04	04	05	06	07	08	09	11
12	01	01	02	04	04	05	05	06	07	08	10	12
13	01	01	03	04	05	05	06	07	08	09	10	13
14	01	01	03	04	05	06	06	07	08	10	11	14
15	01	02	03	05	05	06	07	08	09	11	12	15
16	01	02	03	05	06	06	07	08	10	11	13	16
17	01	02	03	05	06	07	08	09	10	12	14	17
18	01	02	04	05	06	07	08	09	11	13	14	18
19	01	02	04	06	07	08	09	10	11	13	15	19
20	01	02	04	06	07	08	09	10	12	14	16	20
21	01	02	04	06	07	08	09	11	13	15	17	21
22	01	02	04	07	08	09	10	11	13	15	18	22
23	01	02	05	07	08	09	10	12	14	16	18	23
24	01	02	05	07	08	10	11	12	14	17	19	24
25	01	03	05	08	09	10	11	13	15	18	20	25
26	01	03	05	08	09	10	12	13	16	18	21	26
27	01	03	05	08	09	11	12	14	16	19	22	27
28	01	03	06	08	10	11	13	14	17	20	22	28
29	01	03	06	09	10	12	13	15	17	20	23	29
30	02	03	06	09	11	12	14	15	18	21	24	30
33	02	03	07	10	12	13	15	17	20	23	26	33
68	03	06	13	19	22	25	28	32	38	44	50	63
93	05	09	19	28	33	37	42	47	56	65	74	93

37 Dollars.

Years.	½ per ct.	1 per ct.	2 per ct.	3 per ct.	3½ per ct.	4 per ct.	4½ per ct.	5 per ct.	6 per ct.	7 per ct.	8 per ct.	10 per ct.
1	19	37	74	1.11	1.30	1.48	1.67	1.85	2.22	2.59	2.96	3.70
2	37	74	1.48	2.22	2.59	2.96	3.33	3.70	4.44	5.18	5.92	7.40
3	56	1.11	2.22	3.33	3.89	4.44	5.00	5.55	6.66	7.77	8.88	11.10
4	74	1.48	2.96	4.44	5.18	5.92	6.66	7.40	8.88	10.36	11.84	14.80
5	93	1.85	3.70	5.55	6.48	7.40	8.33	9.25	11.10	12.95	14.80	18.50
Months.												
1	02	03	06	09	11	12	14	15	19	22	25	31
2	03	06	12	19	22	25	28	31	37	43	49	62
3	05	09	19	28	32	37	42	46	56	65	74	93
4	06	12	24	37	43	49	56	62	74	86	99	1.23
5	08	15	31	46	54	62	69	77	93	1.08	1.23	1.54
6	09	19	37	56	65	74	83	93	1.11	1.30	1.48	1.85
7	11	22	43	65	76	86	97	1.08	1.30	1.51	1.73	2.16
8	12	25	49	74	86	99	1.11	1.23	1.48	1.73	1.97	2.47
9	14	28	56	83	97	1.11	1.25	1.39	1.67	1.94	2.22	2.78
10	15	31	62	93	1.08	1.23	1.39	1.54	1.85	2.16	2.47	3.08
11	17	36	68	1.02	1.19	1.36	1.53	1.70	2.04	2.37	2.71	3.39
Days.												
1	0	0	0	0	0	0	0	01	01	01	01	01
2	0	0	0	01	01	01	01	01	01	01	02	02
3	0	0	01	01	01	01	01	02	02	02	02	03
4	0	0	01	01	01	02	02	02	02	03	03	04
5	0	01	01	02	02	02	02	03	03	04	04	05
6	0	01	01	02	02	02	03	03	04	04	05	06
7	0	01	01	02	03	03	03	04	04	05	06	07
8	0	01	02	02	03	03	04	04	05	06	07	08
9	0	01	02	03	03	04	04	05	06	06	07	09
10	01	01	02	03	04	04	05	05	06	07	08	10
11	01	01	02	03	04	05	05	06	07	08	09	11
12	01	01	02	04	04	05	06	06	07	09	10	12
13	01	01	03	04	05	05	06	07	08	09	11	13
14	01	01	03	04	05	06	06	07	09	10	12	14
15	01	02	03	05	05	06	07	08	09	11	12	15
16	01	02	03	05	06	07	07	08	10	12	13	16
17	01	02	03	05	06	07	08	09	10	12	14	17
18	01	02	04	06	06	07	08	09	11	13	15	19
19	01	02	04	06	07	08	09	10	12	14	16	20
20	01	02	04	06	07	08	09	10	12	14	16	21
21	01	02	04	06	08	09	10	11	13	15	17	22
22	01	02	05	07	08	09	10	11	14	16	18	23
23	01	02	05	07	08	09	11	12	14	17	19	24
24	01	02	05	07	09	10	11	12	15	17	20	25
25	01	03	05	08	09	10	12	13	15	18	21	26
26	01	03	05	08	09	11	12	13	16	19	21	27
27	01	03	06	08	10	11	12	14	17	19	22	28
28	01	03	06	09	10	12	13	14	17	20	23	29
29	01	03	06	09	10	12	13	15	18	21	24	30
30	02	03	06	09	11	12	14	15	19	22	25	31
33	02	03	07	10	12	14	15	17	20	24	27	34
63	03	06	13	19	23	26	29	32	39	45	52	65
93	05	10	19	29	34	39	43	48	58	68	77	97

38 Dollars.

	½ per ct.	1 per ct.	2 per ct.	3 per ct.	3½ per ct.	4 per ct.	4½ per ct.	5 per ct.	6 per ct.	7 per ct.	8 per ct.	10 per ct.
Years.												
1	19	38	76	1.14	1.33	1.52	1.71	1.90	2.28	2.66	3.04	3.80
2	38	76	1.52	2.28	2.66	3.04	3.42	3.80	4.56	5.32	6.08	7.60
3	57	1.14	2.28	3.42	3.99	4.56	5.13	5.70	6.84	7.98	9.12	11.40
4	76	1.52	3.04	4.56	5.32	6.08	6.84	7.60	9.12	10.64	12.16	15.20
5	95	1.90	3.80	5.70	6.65	7.60	8.53	9.50	11.40	13.30	15.20	19.00
Months.												
1	02	03	06	10	11	13	14	16	19	22	25	32
2	03	06	13	19	22	25	29	32	38	44	51	63
3	05	10	19	29	33	38	43	48	57	67	76	95
4	06	13	25	38	44	51	57	63	76	89	1.01	1.27
5	08	16	32	48	55	63	71	79	95	1.11	1.27	1.58
6	10	19	38	57	67	76	86	95	1.14	1.33	1.52	1.90
7	11	22	44	67	78	89	1.00	1.11	1.33	1.55	1.77	2.22
8	13	25	51	76	89	1.01	1.14	1.27	1.52	1.77	2.03	2.53
9	14	29	57	86	1.00	1.14	1.28	1.43	1.71	2.00	2.28	2.85
10	16	32	63	95	1.11	1.27	1.43	1.58	1.90	2.22	2.53	3.17
11	17	35	70	1.05	1.22	1.39	1.57	1.74	2.09	2.44	2.79	3.48
Days.												
1	0	0	0	0	0	0	0	01	01	01	01	01
2	0	0	0	01	01	01	01	01	01	01	02	02
3	0	0	01	01	01	01	01	02	02	02	03	03
4	0	0	01	01	01	02	02	02	03	03	03	04
5	0	01	01	02	02	02	02	03	03	04	04	05
6	0	01	01	02	02	03	03	03	04	04	05	06
7	0	01	01	02	03	03	03	04	04	05	06	07
8	0	01	02	03	03	03	04	04	05	06	07	08
9	0	01	02	03	03	04	04	05	06	07	08	10
10	01	01	02	03	04	04	05	05	06	07	08	11
11	01	01	02	03	04	05	05	06	07	08	09	12
12	01	01	03	04	04	05	06	06	08	09	10	13
13	01	01	03	04	05	05	06	07	08	10	11	14
14	01	01	03	04	05	06	07	07	09	10	12	15
15	01	02	03	05	06	06	07	08	10	11	13	16
16	01	02	03	05	06	07	08	08	10	12	14	17
17	01	02	04	05	06	07	08	09	11	13	14	18
18	01	02	04	06	07	08	09	10	11	13	15	19
19	01	02	04	06	07	08	09	10	12	14	16	20
20	01	02	04	06	07	08	10	11	13	15	17	21
21	01	02	04	07	08	09	10	11	13	16	18	22
22	01	02	05	07	08	09	10	12	14	16	19	23
23	01	02	05	07	08	10	11	12	15	17	19	24
24	01	03	05	08	09	10	11	13	15	18	20	25
25	01	03	05	08	09	11	12	13	16	18	21	26
26	01	03	05	08	10	11	12	14	16	19	22	27
27	01	03	06	09	10	11	13	14	17	20	23	29
28	01	03	06	09	10	12	13	15	18	21	24	30
29	02	03	06	09	11	12	14	15	18	21	24	31
30	02	03	06	10	11	13	14	16	19	22	25	32
33	02	03	07	10	12	14	16	17	21	24	28	35
63	03	07	13	20	23	27	30	33	40	47	54	67
93	05	10	20	29	34	39	44	49	59	69	79	98

39 Dollars.

	½ per ct.	1 per ct.	2 per ct.	3 per ct.	3½ per ct.	4 per ct.	4½ per ct.	5 per ct.	6 per ct.	7 per ct.	8 per ct.	10 per ct.
Years.												
1	20	39	78	1.17	1.37	1.56	1.76	1.95	2.34	2.73	3.12	3.90
2	39	78	1.56	2.36	2.73	3.12	3.51	3.90	4.68	5.46	6.24	7.80
3	59	1.17	2.34	3.51	4.10	4.68	5.27	5.85	7.02	8.19	9.36	11.70
4	78	1.56	3.12	4.68	5.46	6.24	7.02	7.80	9.36	10.92	12.48	15.60
5	98	1.95	3.90	5.85	6.83	7.80	8.78	9.75	11.70	13.65	15.60	19.50
Months.												
1	02	03	07	10	11	13	15	16	20	23	26	33
2	03	07	13	20	23	26	29	33	39	46	52	65
3	05	10	20	29	34	39	44	49	59	68	78	98
4	07	13	26	39	46	52	59	65	78	91	1.04	1.30
5	08	16	33	49	57	65	73	81	98	1.14	1.30	1.63
6	10	20	39	59	68	78	88	98	1.17	1.37	1.56	1.95
7	11	23	46	68	80	91	1.02	1.14	1.37	1.59	1.82	2.28
8	13	26	52	78	91	1.04	1.17	1.30	1.56	1.82	2.08	2.60
9	15	29	59	88	1.02	1.17	1.32	1.46	1.76	2.05	2.34	2.93
10	16	33	65	98	1.14	1.30	1.46	1.63	1.95	2.28	2.60	3.25
11	18	36	72	1.07	1.25	1.43	1.60	1.79	2.15	2.50	2.86	3.58
Days.												
1	0	0	0	0	0	0	0	01	01	01	01	01
2	0	0	0	01	01	01	01	01	01	02	02	02
3	0	0	01	01	01	01	01	02	02	02	03	03
4	0	0	01	01	02	02	02	02	03	03	03	04
5	0	01	01	02	02	02	02	03	03	04	04	05
6	0	01	01	02	02	03	03	03	04	05	05	07
7	0	01	02	02	03	03	03	04	05	05	06	08
8	0	01	02	03	03	03	04	04	05	06	07	09
9	0	01	02	03	03	04	04	05	06	07	08	10
10	01	01	02	03	04	04	05	05	07	08	09	11
11	01	01	02	04	04	05	05	06	07	08	10	12
12	01	01	03	04	05	05	06	07	08	09	10	13
13	01	01	03	04	05	06	06	07	08	10	11	14
14	01	02	03	05	05	06	07	08	09	11	12	15
15	01	02	03	05	06	07	07	08	10	11	13	16
16	01	02	03	05	06	07	08	09	10	12	14	17
17	01	02	04	06	06	07	08	09	11	13	15	18
18	01	02	04	06	07	08	09	10	12	14	16	20
19	01	02	04	06	07	08	09	10	12	14	16	21
20	01	02	04	07	08	09	10	11	13	15	17	22
21	01	02	05	07	08	09	10	11	14	16	18	23
22	01	02	05	07	08	10	11	12	14	17	19	24
23	01	02	05	07	09	10	11	12	15	17	20	25
24	01	03	05	08	09	10	12	13	16	18	21	26
25	01	03	05	08	09	11	12	14	16	19	22	27
26	01	03	06	08	10	11	13	14	17	20	23	28
27	01	03	06	09	10	12	13	15	18	20	23	29
28	02	03	06	09	11	12	14	15	18	21	24	30
29	02	03	06	09	11	13	14	16	19	22	25	31
30	02	03	07	10	11	13	15	16	20	23	26	33
33	02	04	07	11	13	14	16	18	21	25	29	36
63	03	07	14	20	24	27	31	34	41	48	55	68
93	05	10	20	30	35	40	45	50	60	71	81	1.01

40 Dollars.

Years.	½ per ct.	1 per ct.	2 per ct.	3 per ct.	3½ per ct.	4 per ct.	4½ per ct.	5 per ct.	6 per ct.	7 per ct.	8 per ct.	10 per ct.
1	20	40	80	1.20	1.40	1.60	1.80	2.00	2.40	2.80	3.20	4.00
2	40	80	1.60	2.40	2.80	3.20	3.60	4.00	4.80	5.60	6.40	8.00
3	60	1.20	2.40	3.60	4.20	4.80	5.40	6.00	7.20	8.40	9.60	12.00
4	80	1.60	3.20	4.80	5.60	6.40	7.20	8.00	9.60	11.20	12.80	16.00
5	1.00	2.00	4.00	6.00	7.00	8.00	9.00	10.00	12.00	14.00	16.00	20.00
Months.												
1	02	03	07	10	12	13	15	17	20	23	27	33
2	03	07	13	20	23	27	30	33	40	47	53	67
3	05	10	20	30	35	40	45	50	60	70	80	1.00
4	07	13	27	40	47	53	60	67	80	93	1.07	1.33
5	08	17	33	50	58	67	75	83	1.00	1.17	1.33	1.67
6	10	20	40	60	70	80	90	1.00	1.20	1.40	1.60	2.00
7	12	23	47	70	82	93	1.05	1.17	1.40	1.63	1.87	2.33
8	13	27	53	80	93	1.07	1.20	1.33	1.60	1.87	2.13	2.67
9	15	30	60	90	1.05	1.20	1.35	1.50	1.80	2.10	2.40	3.00
10	17	33	67	1.00	1.17	1.33	1.50	1.67	2.00	2.33	2.67	3.33
11	18	37	73	1.10	1.28	1.47	1.65	1.83	2.20	2.57	2.93	3.67
Days.												
1	0	0	0	0	0	0	01	01	01	01	01	01
2	0	0	0	01	01	01	01	01	01	02	02	02
3	0	0	01	01	01	01	02	02	02	02	03	03
4	0	0	01	01	02	02	02	02	03	03	04	04
5	0	01	01	02	02	02	03	03	03	04	04	06
6	0	01	01	02	02	03	03	03	04	05	05	07
7	0	01	02	02	03	03	04	04	05	05	06	08
8	0	01	02	03	03	04	04	04	05	06	07	09
9	01	01	02	03	04	04	05	05	06	07	08	10
10	01	01	02	03	04	04	05	06	07	08	09	11
11	01	01	02	04	04	05	06	06	07	09	10	12
12	01	01	03	04	05	05	06	07	08	09	11	13
13	01	01	03	04	05	06	07	07	09	10	12	14
14	01	02	03	05	05	06	07	08	09	11	12	16
15	01	02	03	05	06	07	08	08	10	12	13	17
16	01	02	04	05	06	07	08	09	11	12	14	18
17	01	02	04	06	07	08	09	09	11	13	15	19
18	01	02	04	06	07	08	09	10	12	14	16	20
19	01	02	04	06	07	08	10	11	13	15	17	21
20	01	02	04	07	08	09	10	11	13	16	18	22
21	01	02	05	07	08	09	11	12	14	16	19	23
22	01	02	05	07	09	10	11	12	15	17	20	24
23	01	03	05	08	09	10	12	13	15	18	20	26
24	01	03	05	08	09	11	12	13	16	19	21	27
25	01	03	06	08	10	11	13	14	17	19	22	28
26	01	03	06	09	10	12	13	14	17	20	23	29
27	02	03	06	09	11	12	14	15	18	21	24	30
28	02	03	06	09	11	12	14	16	19	22	25	31
29	02	03	06	10	11	13	15	16	19	23	26	32
30	02	03	07	10	12	13	15	17	20	23	27	33
33	02	04	07	11	13	15	17	18	22	26	29	37
63	03	07	14	21	25	28	32	35	42	49	56	70
93	05	10	21	31	36	41	47	52	62	72	83	1.03

41 Dollars.

Years.	½ per ct.	1 per ct.	2 per ct.	3 per ct.	3½ per ct.	4 per ct.	4½ per ct.	5 per ct.	6 per ct.	7 per ct.	8 per ct.	10 per ct.
1	21	41	82	1.23	1.44	1.64	1.85	2.05	2.46	2.87	3.28	4.10
2	41	82	1.64	2.46	2.87	3.28	3.69	4.10	4.92	5.74	6.56	8.20
3	62	1.23	2.46	3.69	4.31	4.92	5.54	6.15	7.38	8.61	9.84	12.30
4	82	1.64	3.28	4.92	5.74	6.56	7.38	8.20	9.84	11.48	13.12	16.40
5	1.03	2.05	4.10	6.15	7.18	8.20	9.23	10.25	12.30	14.35	16.40	20.50
Months.												
1	02	03	07	10	12	14	15	17	21	24	27	34
2	03	07	14	21	24	27	31	34	41	48	55	68
3	05	10	21	31	36	41	46	51	62	72	82	1.03
4	07	14	27	41	48	55	62	68	82	96	1.09	1.37
5	09	17	34	51	60	68	77	85	1.03	1.20	1.37	1.71
6	10	21	41	62	72	82	92	1.03	1.23	1.44	1.64	2.05
7	12	24	48	72	84	96	1.08	1.20	1.44	1.67	1.91	2.39
8	14	27	55	82	96	1.09	1.23	1.37	1.64	1.91	2.19	2.73
9	15	31	62	92	1.08	1.23	1.38	1.54	1.85	2.15	2.46	3.08
10	17	34	68	1.03	1.20	1.37	1.54	1.71	2.05	2.39	2.73	3.42
11	19	38	75	1.13	1.32	1.50	1.69	1.88	2.26	2.63	3.01	3.76
Days.												
1	0	0	0	0	0	0	01	01	01	01	01	01
2	0	0	0	01	01	01	01	01	01	02	02	02
3	0	0	01	01	01	01	02	02	02	02	03	03
4	0	0	01	01	02	02	02	02	03	03	04	05
5	0	01	01	02	02	02	03	03	03	04	05	06
6	0	01	01	02	02	03	03	03	04	05	05	07
7	0	01	02	02	03	03	04	04	05	06	06	08
8	0	01	02	03	03	04	04	05	05	06	07	09
9	01	01	02	03	04	04	05	05	06	07	08	10
10	01	01	02	03	04	05	05	06	07	08	09	11
11	01	01	03	04	04	05	06	06	08	09	10	13
12	01	01	03	04	05	05	06	07	08	10	11	14
13	01	01	03	04	05	06	07	07	09	10	12	15
14	01	02	03	05	06	06	07	08	10	11	13	16
15	01	02	03	05	06	07	08	09	10	12	14	17
16	01	02	04	05	06	07	08	09	11	13	15	18
17	01	02	04	06	07	08	09	10	12	14	15	19
18	01	02	04	06	07	08	09	10	12	14	16	21
19	01	02	04	06	08	09	10	11	13	15	17	22
20	01	02	05	07	08	09	10	11	14	16	18	23
21	01	02	05	07	08	10	11	12	14	17	19	24
22	01	03	05	08	09	10	11	13	15	18	20	25
23	01	03	05	08	09	10	12	13	16	18	21	26
24	01	03	05	08	10	11	12	14	16	19	22	27
25	01	03	06	09	10	11	13	14	17	20	23	28
26	01	03	06	09	10	12	13	15	18	21	24	30
27	02	03	06	09	11	12	14	15	18	22	25	31
28	02	03	06	10	11	13	14	16	19	22	26	32
29	02	03	07	10	12	13	15	17	20	23	26	33
30	02	03	07	10	12	14	15	17	21	24	27	34
33	02	04	08	11	13	15	17	19	23	26	30	38
63	04	07	14	22	25	29	32	36	43	50	57	72
93	05	11	21	32	37	42	48	53	64	74	85	1.06

42 Dollars.

Years.	½ per ct.	1 per ct.	2 per ct.	3 per ct.	3½ per ct.	4 per ct.	4½ per ct.	5 per ct.	6 per ct.	7 per ct.	8 per ct.	10 per ct.
1	21	42	84	1.26	1.47	1.68	1.89	2.10	2.52	2.94	3.36	4.20
2	42	84	1.68	2.52	2.94	3.36	3.78	4.20	5.04	5.88	6.72	8.40
3	63	1.26	2.52	3.78	4.41	5.04	5.67	6.30	7.56	8.82	10.08	12.60
4	84	1.68	3.36	5.04	5.88	6.72	7.56	8.40	10.08	11.76	13.44	16.80
5	1.05	2.10	4.20	6.30	7.35	8.40	9.45	10.50	12.60	14.70	16.80	21.00
Months.												
1	02	04	07	11	12	14	16	18	21	25	28	35
2	04	07	14	21	25	28	32	35	42	49	56	70
3	05	11	21	32	37	42	47	53	63	74	84	1.05
4	07	14	28	42	49	56	63	70	84	98	1.12	1.40
5	09	18	35	53	61	70	79	88	1.05	1.23	1.40	1.75
6	11	21	42	63	74	84	95	1.05	1.26	1.47	1.68	2.10
7	12	25	49	74	86	98	1.10	1.23	1.47	1.72	1.96	2.45
8	14	28	56	84	98	1.12	1.26	1.40	1.68	1.96	2.24	2.80
9	16	32	63	95	1.10	1.26	1.42	1.58	1.89	2.21	2.52	3.15
10	18	35	70	1.05	1.23	1.40	1.58	1.75	2.10	2.45	2.80	3.50
11	19	39	77	1.16	1.35	1.54	1.73	1.93	2.31	2.70	3.08	3.85
Days.												
1	0	0	0	0	0	0	01	01	01	01	01	01
2	0	0	0	01	01	01	01	01	01	02	02	02
3	0	0	01	01	01	01	02	02	02	02	03	04
4	0	0	01	01	02	02	02	02	03	03	04	05
5	0	01	01	02	02	02	03	03	04	04	05	06
6	0	01	01	02	02	03	03	04	04	05	06	07
7	0	01	02	02	03	03	04	04	05	06	07	08
8	0	01	02	03	03	04	04	05	06	07	07	09
9	01	01	02	03	04	04	05	05	06	07	08	11
10	01	01	02	04	04	05	05	06	07	08	09	12
11	01	01	03	04	04	05	06	06	08	09	10	13
12	01	01	03	04	05	06	06	07	08	10	11	14
13	01	02	03	05	05	06	07	08	09	11	12	15
14	01	02	03	05	06	07	07	08	10	11	13	16
15	01	02	04	05	06	07	08	09	11	12	14	18
16	01	02	04	06	07	07	08	09	11	13	15	19
17	01	02	04	06	07	08	09	10	12	14	16	20
18	01	02	04	06	07	08	09	11	13	15	17	21
19	01	02	04	07	08	09	10	11	13	16	18	22
20	01	02	05	07	08	09	11	12	14	16	19	23
21	01	02	05	07	09	10	11	12	15	17	20	24
22	01	03	05	08	09	10	12	13	15	18	21	26
23	01	03	05	08	09	11	12	13	16	19	21	27
24	01	03	06	08	10	11	13	14	17	20	22	28
25	01	03	06	09	10	12	13	15	18	20	23	29
26	02	03	06	09	11	12	14	15	18	21	24	30
27	02	03	06	09	11	13	14	16	19	22	25	32
28	02	03	07	10	11	13	15	16	20	23	26	33
29	02	03	07	10	12	14	15	17	20	24	27	34
30	02	04	07	11	12	14	16	18	21	25	28	35
33	02	04	08	12	13	15	17	19	23	27	31	39
63	04	07	15	22	26	29	33	37	44	51	59	74
93	05	11	22	33	38	43	49	54	65	76	87	1.09

43 Dollars.

Years.	½ per ct.	1 per ct.	2 per ct.	3 per ct.	3½ per ct.	4 per ct.	4½ per ct.	5 per ct.	6 per ct.	7 per ct.	8 per ct.	10 per ct.
1	22	43	86	1.29	1.51	1.72	1.94	2.15	2.58	3.01	3.44	4.30
2	43	86	1.72	2.58	3.01	3.44	3.87	4.30	5.16	6.02	6.88	8.60
3	65	1.29	2.58	3.87	4.52	5.16	5.81	6.45	7.74	9.03	10.32	12.90
4	86	1.72	3.44	5.16	6.02	6.88	7.74	8.60	10.32	12.04	13.76	17.20
5	1.08	2.15	4.30	6.47	7.53	8.60	9.68	10.75	12.90	15.05	17.20	21.50
Months.												
1	02	04	07	11	13	14	16	18	22	25	29	36
2	04	07	14	22	25	29	32	36	43	50	57	72
3	05	11	22	32	38	43	48	54	65	75	86	1.08
4	07	14	29	43	50	57	65	72	86	1.00	1.15	1.43
5	09	18	36	54	63	72	81	90	1.08	1.25	1.43	1.79
6	11	22	43	65	75	86	97	1.08	1.29	1.51	1.72	2.15
7	13	25	50	75	88	1.00	1.13	1.25	1.51	1.76	2.01	2.51
8	14	29	57	86	1.00	1.15	1.29	1.43	1.72	2.01	2.29	2.87
9	16	32	65	97	1.13	1.29	1.45	1.61	1.94	2.26	2.58	3.23
10	18	36	72	1.08	1.25	1.43	1.61	1.79	2.15	2.51	2.87	3.58
11	20	39	79	1.18	1.38	1.58	1.77	1.97	2.37	2.76	3.15	3.94
Days.												
1	0	0	0	0	0	0	01	01	01	01	01	01
2	0	0	0	01	01	01	01	01	01	02	02	02
3	0	0	01	01	01	01	02	02	02	03	03	04
4	0	0	01	01	02	02	02	02	03	03	04	05
5	0	01	01	02	02	02	03	03	04	04	05	06
6	0	01	01	02	03	03	03	04	04	05	06	07
7	0	01	02	03	03	03	04	04	05	06	07	08
8	0	01	02	03	03	04	04	05	06	07	08	10
9	01	01	02	03	04	04	05	05	06	08	09	11
10	01	01	02	04	04	05	05	06	07	08	10	12
11	01	01	03	04	05	05	06	07	08	09	11	13
12	01	01	03	04	05	06	06	07	09	10	11	14
13	01	02	03	05	05	06	07	08	09	11	12	16
14	01	02	03	05	06	07	08	08	10	12	13	17
15	01	02	04	05	06	07	08	09	11	13	14	18
16	01	02	04	06	07	08	09	10	11	13	15	19
17	01	02	04	06	07	08	09	10	12	14	16	20
18	01	02	04	06	08	09	10	11	13	15	17	22
19	01	02	05	07	08	09	10	11	14	16	18	23
20	01	02	05	07	08	10	11	12	14	17	19	24
21	01	03	05	08	09	10	11	13	15	18	20	25
22	01	03	05	08	09	11	12	13	16	18	21	26
23	01	03	05	08	10	11	12	14	16	19	22	27
24	01	03	06	09	10	11	13	14	17	20	23	29
25	01	03	06	09	10	12	13	15	18	21	24	30
26	02	03	06	09	11	12	14	16	19	22	25	31
27	02	03	06	10	11	13	15	16	19	23	26	32
28	02	03	07	10	12	13	15	17	20	23	27	33
29	02	03	07	10	12	14	16	17	21	24	28	35
30	02	04	07	11	13	14	16	18	22	25	29	36
33	02	04	08	12	14	16	18	20	24	28	32	39
63	04	08	15	23	26	30	34	38	45	53	60	75
93	06	11	22	33	39	44	50	56	67	78	89	1.11

44 Dollars.

Years.	½ per ct.	1 per ct.	2 per ct.	3 per ct.	3½ per ct.	4 per ct.	4½ per ct.	5 per ct.	6 per ct.	7 per ct.	8 per ct.	10 per ct.
1	22	44	88	1.32	1.54	1.76	1.98	2.20	2.64	3.08	3.52	4.40
2	44	88	1.76	2.64	3.08	3.52	3.96	4.40	5.28	6.16	7.04	8.80
3	66	1.32	2.64	3.96	4.62	5.28	5.94	6.60	7.92	9.24	10.56	13.20
4	88	1.76	3.52	5.28	6.16	7.04	7.92	8.80	10.56	12.32	14.08	17.60
5	1.10	2.20	4.40	6.60	7.70	8.80	9.90	11.00	13.20	15.40	17.60	22.00
Months.												
1	02	04	07	11	13	15	17	18	22	26	29	37
2	04	07	15	22	26	29	33	37	44	51	59	73
3	06	11	22	33	39	44	50	55	66	77	88	1.10
4	07	15	29	44	51	59	66	73	88	1.03	1.17	1.47
5	09	18	37	55	64	73	83	92	1.10	1.28	1.47	1.83
6	11	22	44	66	77	88	99	1.10	1.32	1.54	1.76	2.20
7	13	26	51	77	90	1.03	1.16	1.28	1.54	1.80	2.05	2.57
8	15	29	59	88	1.03	1.17	1.32	1.47	1.76	2.05	2.35	2.93
9	17	33	66	99	1.16	1.32	1.49	1.65	1.98	2.31	2.64	3.30
10	18	37	73	1.10	1.28	1.47	1.65	1.83	2.20	2.57	2.93	3.67
11	20	40	81	1.21	1.41	1.61	1.82	2.02	2.42	2.82	3.23	4.03
Days.												
1	0	0	0	0	0	0	01	01	01	01	01	01
2	0	0	0	01	01	01	01	01	01	02	02	02
3	0	0	01	01	01	01	02	02	02	03	03	04
4	0	0	01	01	02	02	02	02	03	03	04	05
5	0	01	01	02	02	02	03	03	04	04	05	06
6	0	01	01	02	03	03	03	04	04	05	06	07
7	0	01	02	03	03	03	04	04	05	06	07	09
8	0	01	02	03	03	04	04	05	06	07	08	10
9	01	01	02	03	04	04	05	06	07	08	09	11
10	01	01	02	04	04	05	06	06	07	09	10	12
11	01	01	03	04	05	05	06	07	08	09	11	13
12	01	01	03	04	05	06	07	07	09	10	12	15
13	01	02	03	05	06	06	07	08	10	11	13	16
14	01	02	03	05	06	07	08	09	10	12	14	17
15	01	02	04	06	06	07	08	09	11	13	15	18
16	01	02	04	06	07	08	09	10	12	14	16	20
17	01	02	04	06	07	08	09	10	12	15	17	21
18	01	02	04	07	08	09	10	11	13	15	18	22
19	01	02	05	07	08	09	10	12	14	16	19	23
20	01	02	05	07	09	10	11	12	15	17	20	24
21	01	03	05	08	09	10	12	13	15	18	21	26
22	01	03	05	08	09	11	12	13	16	19	22	27
23	01	03	06	08	10	11	13	14	17	20	22	28
24	01	03	06	09	10	12	13	15	18	21	23	29
25	02	03	06	09	11	12	14	15	18	21	24	31
26	02	03	06	10	11	13	14	16	19	22	25	32
27	02	03	07	10	12	13	15	17	20	23	26	33
28	02	03	07	10	12	14	15	17	21	24	27	34
29	02	04	07	11	12	14	16	18	21	25	28	35
30	02	04	07	11	13	15	17	18	22	26	29	37
33	02	04	08	12	14	16	18	20	24	28	32	40
63	04	08	15	23	27	31	35	39	46	54	62	77
93	06	11	23	34	40	45	51	57	68	80	91	1.14

45 Dollars.

Years.	½ per ct.	1 per ct.	2 per ct.	3 per ct.	3½ per ct.	4 per ct.	4½ per ct.	5 per ct.	6 per ot.	7 per ct.	8 per ct.	10 per ct.
1	23	45	90	1.35	1.58	1.80	2.03	2.25	2.70	3.15	3.60	4.50
2	45	90	1.80	2.70	3.15	3.60	4.05	4.50	5.40	6.30	7.20	9.00
3	68	1.35	2.70	4.05	5.73	5.40	6.08	6.75	8.10	9.45	10.80	13.50
4	90	1.80	3.60	5.40	6.30	7.20	8.10	9.00	10.80	12.60	14.40	18.00
5	1.13	2.25	4.50	6.75	7.88	9.00	10.13	11.25	13.50	15.75	18.00	22.50
Months.												
1	02	04	08	11	13	15	17	19	23	26	30	38
2	04	08	15	23	26	30	34	38	45	53	60	75
3	06	11	23	34	39	45	51	56	68	79	90	1.13
4	08	15	30	45	53	60	68	75	90	1.05	1.20	1.50
5	09	19	38	56	66	75	84	94	1.13	1.31	1.50	1.88
6	11	23	45	68	79	90	1.01	1.13	1.35	1.58	1.80	2.25
7	13	26	53	79	92	1.05	1.18	1.31	1.58	1.84	2.10	2.62
8	15	30	60	90	1.05	1.20	1.35	1.50	1.80	2.10	2.40	3.00
9	17	34	68	1.01	1.18	1.35	1.52	1.69	2.03	2.36	2.70	3.38
10	19	38	75	1.13	1.31	1.50	1.69	1.88	2.25	2.63	3.00	3.75
11	21	41	83	1.24	1.44	1.65	1.86	2.06	2.48	2.89	3.30	4.13
Days.												
1	0	0	0	0	0	01	01	01	01	01	01	01
2	0	0	01	01	01	01	01	01	02	02	02	03
3	0	0	01	01	01	02	02	02	02	03	03	04
4	0	01	01	02	02	02	02	03	03	04	04	05
5	0	01	01	02	02	03	03	03	04	04	05	06
6	0	01	02	02	03	03	03	04	05	05	06	08
7	0	01	02	03	03	04	04	04	05	06	07	09
8	0	01	02	03	04	04	05	05	06	07	08	10
9	01	01	02	03	04	05	05	06	07	08	09	11
10	01	01	03	04	04	05	06	06	08	09	10	13
11	01	01	03	04	05	06	06	07	08	10	11	14
12	01	02	03	05	05	06	07	08	09	11	12	15
13	01	02	03	05	06	07	07	08	10	11	13	16
14	01	02	04	05	06	07	08	09	11	12	14	18
15	01	02	04	06	07	08	08	09	11	13	15	19
16	01	02	04	06	07	08	09	10	12	14	16	20
17	01	02	04	06	07	09	10	11	13	15	17	21
18	01	02	05	07	08	09	10	11	14	16	18	23
19	01	02	05	07	08	10	11	12	14	17	19	24
20	01	03	05	08	09	10	11	13	15	18	20	25
21	01	03	05	08	09	11	12	13	16	18	21	26
22	01	03	06	08	10	11	12	14	17	19	22	28
23	01	03	06	09	10	12	13	14	17	20	23	29
24	01	03	06	09	11	12	14	15	18	21	24	30
25	02	03	06	09	11	13	14	16	19	22	25	31
26	02	03	07	10	11	13	15	16	20	23	26	33
27	02	03	07	10	12	14	15	17	20	24	27	34
28	02	04	07	11	12	14	16	18	21	25	28	35
29	02	04	07	11	13	15	16	18	22	25	29	36
30	02	04	08	11	13	15	17	19	23	26	30	38
33	02	04	08	12	14	17	19	21	25	29	33	41
63	04	08	16	24	28	32	35	39	47	55	63	79
93	06	11	23	35	41	47	52	58	70	81	93	1.16

46 Dollars.

	½ per ct.	1 per ct.	2 per ct.	3 per ct.	3½ per ct.	4 per ct.	4½ per ct.	5 per ct.	6 per ct.	7 per ct.	8 per ct.	10 per ct.
Years.												
1	23	46	92	1.38	1.61	1.84	2.07	2.30	2.76	3.22	3.68	4.60
2	46	92	1.84	2.76	3.22	3.68	4.14	4.60	5.52	6.44	7.36	9.20
3	69	1.38	2.76	4.14	4.83	5.52	6.21	6.90	8.28	9.66	11.04	13.80
4	92	1.84	3.68	5.52	6.44	7.36	8.28	9.20	11.04	12.88	14.72	18.40
5	1.15	2.30	4.60	6.90	8.05	9.20	10.35	11.50	13.80	16.10	18.40	23.00
Months.												
1	02	04	08	12	13	15	17	19	23	27	31	38
2	04	08	15	23	27	31	35	38	46	54	61	77
3	06	12	23	35	40	46	52	58	69	81	92	1.15
4	08	15	31	46	54	61	69	77	92	1.07	1.23	1.53
5	10	19	38	58	67	77	86	96	1.15	1.34	1.53	1.92
6	12	23	46	69	81	92	1.04	1.15	1.38	1.61	1.84	2.30
7	13	27	54	81	94	1.07	1.21	1.34	1.61	1.88	2.15	2.68
8	15	31	61	92	1.07	1.23	1.38	1.53	1.84	2.15	2.45	3.07
9	17	35	69	1.04	1.21	1.38	1.55	1.73	2.07	2.42	2.76	3.45
10	19	38	77	1.15	1.34	1.53	1.73	1.92	2.30	2.68	3.07	3.83
11	21	42	84	1.27	1.48	1.69	1.90	2.11	2.53	2.95	3.37	4.22
Days.												
1	0	0	0	0	0	01	01	01	01	01	01	01
2	0	0	01	01	01	01	01	01	02	02	02	03
3	0	0	01	01	01	02	02	02	02	03	03	04
4	0	01	01	02	02	02	02	03	03	04	04	05
5	0	01	01	02	02	03	03	03	04	04	05	06
6	0	01	02	02	03	03	03	04	05	05	06	08
7	0	01	02	03	03	04	04	04	05	06	07	09
8	0	01	02	03	04	04	05	05	06	07	08	10
9	01	01	02	03	04	05	05	06	07	08	09	12
10	01	01	03	04	04	05	06	06	08	09	10	13
11	01	01	03	04	05	06	06	07	08	10	11	14
12	01	02	03	05	05	06	07	08	09	11	12	15
13	01	02	03	05	06	07	07	08	10	12	13	17
14	01	02	04	05	06	07	08	09	11	13	14	18
15	01	02	04	06	07	08	09	10	12	13	15	19
16	01	02	04	06	07	08	10	10	12	14	16	20
17	01	02	04	07	08	09	10	11	13	15	17	22
18	01	02	05	07	08	09	11	12	14	16	18	23
19	01	02	05	07	08	10	11	12	15	17	19	24
20	01	03	05	08	09	10	12	13	15	18	20	26
21	01	03	05	08	09	11	12	13	16	19	21	27
22	01	03	06	08	10	11	13	14	17	20	22	28
23	01	03	06	09	10	12	13	15	18	21	24	29
24	02	03	06	09	11	12	14	15	18	21	25	31
25	02	03	06	10	11	13	14	16	19	22	26	32
26	02	03	07	10	12	13	15	17	20	23	27	33
27	02	03	07	10	12	14	16	17	21	24	28	35
28	02	04	07	11	13	14	16	18	21	25	29	36
29	02	04	07	11	13	15	17	19	22	26	30	37
30	02	04	08	12	13	15	17	19	23	27	31	38
33	02	04	08	13	15	17	19	21	25	30	34	42
63	04	08	16	24	28	32	36	40	48	56	64	81
93	06	12	24	36	41	48	53	59	71	83	95	1.19

47 Dollars.

Years.	½ per ct.	1 per ct.	2 per ct.	3 per ct.	3½ per ct.	4 per ct.	4½ per ct.	5 per ct.	6 per ct.	7 per ct.	8 per ct.	10 per ct.
1	24	47	94	1.41	1.65	1.88	2.12	2.35	2.82	3.29	3.76	4.70
2	47	94	1.88	2.82	3.29	3.76	4.23	4.70	5.64	6.58	7.52	9.40
3	71	1.41	2.82	4.23	4.94	5.64	6.35	7.05	8.46	9.87	11.28	14.10
4	94	1.88	3.76	5.64	6.58	7.52	8.46	9.40	11.28	13.16	15.04	18.80
5	1.18	2.35	4.70	7.05	8.23	9.40	10.58	11.75	14.10	16.45	18.80	23.50
Months.												
1	02	04	08	12	14	16	18	20	24	27	31	39
2	04	08	16	24	27	31	35	39	47	55	63	78
3	06	12	24	35	41	47	53	59	71	82	94	1.18
4	08	16	31	47	55	63	71	78	94	1.10	1.25	1.57
5	10	20	39	59	69	78	88	98	1.18	1.37	1.57	1.96
6	12	24	47	71	82	94	1.06	1.18	1.41	1.65	1.88	2.35
7	14	27	55	82	96	1.10	1.23	1.37	1.65	1.92	2.19	2.74
8	16	31	63	94	1.10	1.25	1.41	1.57	1.88	2.19	2.51	3.13
9	18	35	71	1.06	1.23	1.41	1.59	1.76	2.12	2.47	2.82	3.53
10	20	39	78	1.18	1.37	1.57	1.76	1.96	2.35	2.74	3.13	3.92
11	22	43	86	1.29	1.51	1.72	1.94	2.15	2.59	3.02	3.45	4.31
Days.												
1	0	0	0	0	0	01	01	01	01	01	01	01
2	0	0	01	01	01	01	01	01	02	02	02	03
3	0	0	01	01	01	02	02	02	02	03	03	04
4	0	01	01	02	02	02	02	03	03	04	04	05
5	0	01	01	02	02	03	03	03	04	05	05	07
6	0	01	02	02	03	03	04	04	05	05	06	08
7	0	01	02	03	03	04	04	05	05	06	07	09
8	01	01	02	03	04	04	05	05	06	07	08	10
9	01	01	02	04	04	05	05	06	07	08	09	12
10	01	01	03	04	05	05	06	07	08	09	10	13
11	01	01	03	04	05	06	06	07	09	10	11	14
12	01	02	03	05	05	06	07	08	09	11	13	16
13	01	02	03	05	06	07	08	08	10	12	14	17
14	01	02	04	05	06	07	08	09	11	13	15	18
15	01	02	04	06	07	08	09	10	12	14	16	20
16	01	02	04	06	07	08	09	10	13	15	17	21
17	01	02	04	07	08	09	10	11	13	16	18	22
18	01	02	05	07	08	09	11	12	14	16	19	24
19	01	02	05	07	09	10	11	12	15	17	20	25
20	01	03	05	08	09	10	12	13	16	18	21	26
21	01	03	05	08	10	11	12	14	16	19	22	27
22	01	03	06	09	10	11	13	14	17	20	23	29
23	02	03	06	09	11	12	14	15	18	21	24	30
24	02	03	06	09	11	13	14	16	19	22	25	31
25	02	03	07	10	11	13	15	16	20	23	26	33
26	02	03	07	10	12	14	15	17	20	24	27	34
27	02	04	07	11	12	14	16	18	21	25	28	35
28	02	04	07	11	13	15	16	18	22	26	29	37
29	02	04	08	11	13	15	17	19	23	27	30	38
30	02	04	08	12	14	16	18	20	24	27	31	39
33	02	04	09	13	15	17	19	22	26	30	34	43
63	04	08	16	25	29	33	37	41	49	58	66	82
93	06	12	24	36	42	49	55	61	73	85	97	1.21

48 Dollars.

	½ per ct.	1 per ct.	2 per ct.	3 per ct.	3½ per ct.	4 per ct.	4½ per ct.	5 per ct.	6 per ct.	7 per ct.	8 per ct.	10 per ct.
Years.												
1	24	48	96	1.44	1.68	1.92	2.16	2.40	2.88	3.36	3.84	4.80
2	48	96	1.92	2.88	3.36	3.84	4.32	4.80	5.76	6.72	7.68	9.60
3	72	1.44	2.88	4.32	5.04	5.76	6.48	7.20	8.64	10.08	11.52	14.40
4	96	1.92	3.84	5.76	6.72	7.68	8.64	9.60	11.52	13.44	15.36	19.20
5	1.20	2.40	4.80	7.20	8.40	9.60	10.80	12.00	14.40	16.80	19.20	24.00
Months.												
1	02	04	08	12	14	16	18	20	24	28	32	40
2	04	08	16	24	28	32	36	40	48	56	64	80
3	06	12	24	36	42	48	54	60	72	84	96	1.20
4	08	16	32	48	56	64	72	80	96	1.12	1.28	1.60
5	10	20	40	60	70	80	90	1.00	1.20	1.40	1.60	2.00
6	12	24	48	72	84	96	1.08	1.20	1.44	1.68	1.92	2.40
7	14	28	56	84	98	1.12	1.26	1.40	1.68	1.96	2.24	2.80
8	16	32	64	96	1.12	1.28	1.44	1.60	1.92	2.24	2.56	3.20
9	18	36	72	1.08	1.26	1.44	1.62	1.80	2.16	2.52	2.88	3.60
10	20	40	80	1.20	1.40	1.60	1.80	2.00	2.40	2.80	3.20	4.00
11	22	44	88	1.32	1.54	1.76	1.98	2.20	2.64	3.08	3.52	4.40
Days.												
1	0	0	0	0	0	01	01	01	01	01	01	01
2	0	0	01	01	01	01	01	01	02	02	02	03
3	0	0	01	01	01	02	02	02	02	03	03	04
4	0	01	01	02	02	02	02	03	03	04	04	05
5	0	01	01	02	02	03	03	03	04	05	05	07
6	0	01	02	02	03	03	04	04	05	06	06	08
7	0	01	02	03	03	04	04	05	06	07	07	09
8	01	01	02	03	04	04	05	05	06	07	09	11
9	01	01	02	04	04	05	05	06	07	08	10	12
10	01	01	03	04	05	05	06	07	08	09	11	13
11	01	01	03	04	05	06	07	07	09	10	12	15
12	01	02	03	05	06	06	07	08	10	11	13	16
13	01	02	03	05	06	07	08	09	10	12	14	17
14	01	02	04	06	07	07	08	09	11	13	15	19
15	01	02	04	06	07	08	09	10	12	14	16	20
16	01	02	04	06	07	09	10	11	13	15	17	21
17	01	02	05	07	08	09	10	11	14	16	18	23
18	01	02	05	07	08	10	11	12	14	17	19	24
19	01	03	05	08	09	10	11	13	15	18	20	25
20	01	03	05	08	09	11	12	13	16	19	21	27
21	01	03	06	08	10	11	13	14	17	20	22	28
22	01	03	06	09	10	12	13	15	18	21	23	29
23	02	03	06	09	11	12	14	15	18	21	25	31
24	02	03	06	10	11	13	14	16	19	22	26	32
25	02	03	07	10	12	13	15	17	20	23	27	33
26	02	03	07	10	12	14	16	17	21	24	28	35
27	02	04	07	11	13	14	16	18	22	25	29	36
28	02	04	07	11	13	15	17	19	22	26	30	37
29	02	04	08	12	14	15	17	19	23	27	31	39
30	02	04	08	12	14	16	18	20	24	28	32	40
33	02	04	09	13	15	18	20	22	26	31	35	44
63	04	08	17	25	29	34	38	42	50	59	67	84
93	06	12	25	37	43	50	56	62	74	87	99	1.24

49 Dollars.

	½ per ct.	1 per ct.	2 per ct.	3 per ct.	3½ per ct.	4 per ct.	4½ per ct.	5 per ct.	6 per ct.	7 per ct.	8 per ct.	10 per ct.
Years.												
1	25	49	98	1.47	1.72	1.96	2.21	2.45	2.94	3.43	3.92	4.90
2	49	98	1.96	2.94	3.43	3.92	4.41	4.90	5.88	6.86	7.84	9.80
3	74	1.47	2.94	4.41	5.15	5.88	6.62	7.35	8.82	10.29	11.76	14.70
4	98	1.96	3.92	5.88	6.86	7.84	8.82	9.80	11.76	13.72	15.68	19.60
5	1.23	2.45	4.90	7.35	8.58	9.80	11.03	12.25	14.70	17.15	19.60	24.50
Months.												
1	02	04	08	12	14	16	18	20	25	29	33	41
2	04	08	16	25	29	33	37	41	49	57	65	82
3	06	12	25	37	43	49	55	61	74	86	98	1.23
4	08	16	33	49	57	65	74	82	98	1.14	1.31	1.63
5	10	20	41	61	71	82	92	1.02	1.23	1.43	1.63	2.04
6	12	25	49	74	86	98	1.10	1.23	1.47	1.72	1.96	2.45
7	14	29	57	86	1.00	1.14	1.29	1.43	1.72	2.00	2.29	2.86
8	16	33	65	98	1.14	1.31	1.47	1.63	1.96	2.29	2.61	3.27
9	18	37	74	1.10	1.29	1.47	1.65	1.84	2.21	2.57	2.94	3.68
10	20	41	82	1.23	1.43	1.63	1.84	2.04	2.45	2.86	3.27	4.08
11	22	45	90	1.35	1.57	1.80	2.02	2.25	2.70	3.14	3.59	4.49
Days.												
1	0	0	0	0	0	01	01	01	01	01	01	01
2	0	0	01	01	01	01	01	01	02	02	02	03
3	0	0	01	01	01	02	02	02	02	03	03	04
4	0	01	01	02	02	02	02	03	03	04	04	05
5	0	01	01	02	02	03	03	03	04	05	05	07
6	0	01	02	02	03	03	04	04	05	06	07	08
7	0	01	02	03	03	04	04	05	06	07	08	10
8	01	01	02	03	04	04	05	05	07	08	09	11
9	01	01	02	04	04	05	06	06	07	09	10	12
10	01	01	03	04	05	05	06	07	08	10	11	14
11	01	01	03	04	05	06	07	07	09	10	12	15
12	01	02	03	05	06	07	07	08	10	11	13	16
13	01	02	04	05	06	07	08	09	11	12	14	18
14	01	02	04	06	07	08	09	10	11	13	15	19
15	01	02	04	06	07	08	09	10	12	14	16	20
16	01	02	04	07	08	09	10	11	13	15	17	22
17	01	02	05	07	08	09	10	12	14	16	19	23
18	01	02	05	07	09	10	11	12	15	17	20	25
19	01	03	05	08	09	10	12	13	16	18	21	26
20	01	03	05	08	10	11	12	14	16	19	22	27
21	01	03	06	09	10	11	13	14	17	20	23	29
22	01	03	06	09	10	12	13	15	18	21	24	30
23	02	03	06	09	11	13	14	16	19	22	25	31
24	02	03	07	10	11	13	15	16	20	23	26	33
25	02	03	07	10	12	14	15	17	20	24	27	34
26	02	04	07	11	12	14	16	18	21	25	28	35
27	02	04	07	11	13	15	17	18	22	26	29	37
28	02	04	08	11	13	15	17	19	23	27	30	38
29	02	04	08	12	14	16	18	20	24	28	32	39
30	02	04	08	12	14	16	18	20	25	29	33	41
33	02	04	09	13	16	18	20	22	27	31	36	45
63	04	09	17	26	30	34	39	43	51	60	69	86
93	06	13	25	38	44	51	57	63	76	89	1.01	1.27

50 Dollars.

Years.	½ per ct.	1 per ct.	2 per ct.	3 per ct.	3½ per ct.	4 per ct.	4½ per ct.	5 per ct.	6 per ct.	7 per ct.	8 per ct.	10 per ct.
1	25	50	1.00	1.50	1.75	2.00	2.25	2.50	3.00	3.50	4.00	5.00
2	50	1.00	2.00	3.00	3.50	4.00	4.50	5.00	6.00	7.00	8.00	10.00
3	75	1.50	3.00	4.50	5.25	6.00	6.75	7.50	9.00	10.50	12.00	15.00
4	1.00	2.00	4.00	6.00	7.00	8.00	9.00	10.00	12.00	14.00	16.00	20.00
5	1.25	2.50	5.00	7.50	8.75	10.00	11.25	12.50	15.00	17.50	20.00	25.00
Months.												
1	02	04	08	13	15	17	19	21	25	29	33	42
2	04	08	17	25	29	33	38	42	50	58	67	83
3	06	13	25	38	44	50	56	63	75	88	1.00	1.25
4	08	17	33	50	58	67	75	83	1.00	1.17	1.33	1.67
5	10	21	42	63	73	83	94	1.04	1.25	1.46	1.67	2.08
6	13	25	50	75	88	1.00	1.13	1.25	1.50	1.75	2.00	2.50
7	15	29	58	88	1.02	1.17	1.31	1.46	1.75	2.04	2.33	2.92
8	17	33	67	1.00	1.17	1.33	1.50	1.67	2.00	2.33	2.67	3.33
9	19	38	75	1.13	1.31	1.50	1.69	1.88	2.25	2.63	3.00	3.75
10	21	42	83	1.25	1.46	1.67	1.88	2.08	2.50	2.92	3.33	4.17
11	23	46	92	1.38	1.60	1.83	2.06	2.29	2.75	3.21	3.67	4.58
Days.												
1	0	0	0	0	0	01	01	01	01	01	01	01
2	0	0	01	01	01	01	01	01	02	02	02	03
3	0	0	01	01	01	02	02	02	03	03	03	04
4	0	01	01	02	02	02	03	03	03	04	04	06
5	0	01	01	02	02	03	03	03	04	05	06	07
6	0	01	02	03	03	03	04	04	05	06	07	08
7	0	01	02	03	03	04	04	05	06	07	08	10
8	01	01	02	03	04	04	05	06	07	08	09	11
9	01	01	03	04	04	05	06	06	08	09	10	13
10	01	01	03	04	05	06	06	07	08	10	11	14
11	01	02	03	05	05	06	07	08	09	11	12	15
12	01	02	03	05	06	07	08	08	10	12	13	17
13	01	02	04	05	06	07	08	09	11	13	14	18
14	01	02	04	06	07	08	09	10	12	14	16	19
15	01	02	04	06	07	08	09	10	13	15	17	21
16	01	02	04	07	08	09	10	11	13	16	18	22
17	01	02	05	07	08	09	11	12	14	17	19	24
18	01	03	05	08	09	10	11	13	15	18	20	25
19	01	03	05	08	09	11	12	13	16	18	21	26
20	01	03	06	08	10	11	13	14	17	19	22	28
21	01	03	06	09	10	12	13	15	18	20	23	29
22	02	03	06	09	11	12	14	15	18	21	24	31
23	02	03	06	10	11	13	14	16	19	22	26	32
24	02	03	07	10	12	13	15	17	20	23	27	33
25	02	03	07	10	12	14	16	17	21	24	28	35
26	02	04	07	11	13	14	16	18	22	25	29	36
27	02	04	08	11	13	15	17	19	23	26	30	38
28	02	04	08	12	14	16	18	19	23	27	31	39
29	02	04	08	12	14	16	18	20	24	28	32	40
30	02	04	08	13	15	17	19	21	25	29	33	42
33	02	05	09	14	16	18	21	23	28	32	37	46
63	04	09	18	26	31	35	39	44	53	61	70	88
93	06	13	26	39	45	52	58	65	78	90	1.03	1.29

51 Dollars.

Years.	½ per ct.	1 per ct.	2 per ct.	3 per ct.	3½ per ct.	4 per ct.	4½ per ct.	5 per ct.	6 per ct.	7 per ct.	8 per ct.	10 per ct.
1	26	51	1.02	1.53	1.79	2.04	2.30	2.55	3.06	3.57	4.08	5.10
2	51	1.02	2.04	3.06	3.57	4.08	4.59	5.10	6.12	7.14	8.16	10.20
3	77	1.53	3.06	4.59	5.36	6.12	6.89	7.65	9.18	10.71	12.24	15.30
4	1.02	2.04	4.08	6.12	7.14	8.16	9.18	10.20	12.24	14.28	16.32	20.40
5	1.28	2.55	5.10	7.65	8.93	10.20	11.48	12.75	15.30	17.85	20.40	25.50
Months.												
1	02	04	09	13	15	17	19	21	26	30	34	43
2	04	09	17	26	30	34	38	43	51	60	68	85
3	06	13	26	38	45	51	57	64	77	89	1.02	1.28
4	09	17	34	51	60	68	77	85	1.02	1.19	1.36	1.70
5	11	21	43	64	74	85	96	1.06	1.28	1.49	1.70	2.13
6	13	26	51	77	89	1.02	1.15	1.28	1.53	1.79	2.04	2.55
7	15	30	60	89	1.04	1.19	1.34	1.49	1.79	2.08	2.38	2.98
8	17	34	68	1.02	1.19	1.36	1.53	1.70	2.04	2.38	2.72	3.40
9	19	38	77	1.15	1.34	1.53	1.72	1.91	2.30	2.68	3.06	3.83
10	21	43	85	1.28	1.49	1.70	1.91	2.13	2.55	2.98	3.40	4.25
11	23	47	94	1.40	1.64	1.87	2.10	2.34	2.81	3.27	3.74	4.68
Days.												
1	0	0	0	0	01	01	01	01	01	01	01	01
2	0	0	01	01	01	01	01	01	02	02	02	03
3	0	0	01	01	01	02	02	02	03	03	03	04
4	0	01	01	02	02	02	03	03	03	04	05	06
5	0	01	01	02	02	03	03	04	04	05	06	07
6	0	01	02	03	03	03	04	04	05	06	07	09
7	01	01	02	03	03	04	04	05	06	07	08	10
8	01	01	02	03	04	05	05	06	07	08	09	11
9	01	01	03	04	04	05	06	06	08	09	10	13
10	01	01	03	04	05	06	06	07	09	10	11	14
11	01	02	03	05	05	06	07	08	09	11	12	16
12	01	02	03	05	06	07	08	09	10	12	14	17
13	01	02	04	06	06	07	08	09	11	13	15	18
14	01	02	04	06	07	08	09	10	12	14	16	20
15	01	02	04	06	07	09	10	11	13	15	17	21
16	01	02	05	07	08	09	10	11	14	16	18	23
17	01	02	05	07	08	10	11	12	14	17	19	24
18	01	03	05	08	09	10	11	13	15	18	20	26
19	01	03	05	08	09	11	12	13	16	19	22	27
20	01	03	06	09	10	11	13	14	17	20	23	28
21	01	03	06	09	10	12	13	15	18	21	24	30
22	02	03	06	09	11	12	14	16	19	22	25	31
23	02	03	07	10	11	13	15	16	20	23	26	33
24	02	03	07	10	12	14	15	17	20	24	27	34
25	02	04	07	11	12	14	16	18	21	25	28	35
26	02	04	07	11	13	15	17	18	22	26	29	37
27	02	04	08	11	13	15	17	19	23	27	31	38
28	02	04	08	12	14	16	18	20	24	28	32	40
29	02	04	08	12	14	16	18	21	25	29	33	41
30	02	04	09	13	15	17	19	21	26	30	34	43
33	02	05	09	14	16	19	21	23	28	33	37	47
63	04	09	18	27	31	36	40	45	54	62	71	89
93	07	13	26	40	46	53	59	66	79	92	1.05	1.32

52 Dollars.

Years.	½ per ct.	1 per ct.	2 per ct.	3 per ct.	3½ per ct.	4 per ct.	4½ per ct.	5 per ct.	6 per ct.	7 per ct.	8 per ct.	10 per ct.
1	26	52	1.04	1.56	1.82	2.08	2.34	2.60	3.12	3.64	4.16	5.20
2	52	1.04	2.08	3.12	3.64	4.16	4.68	5.20	6.24	7.28	8.32	10.40
3	78	1.56	3.12	4.68	5.46	6.24	7.02	7.80	9.36	10.92	12.48	15.60
4	1.04	2.08	4.16	6.24	7.28	8.32	9.36	10.40	12.48	14.56	16.64	20.80
5	1.30	2.60	5.20	7.80	9.10	10.40	11.70	13.00	15.60	18.20	20.80	26.00
Months.												
1	02	04	09	13	15	17	20	22	26	30	35	43
2	04	09	17	26	30	35	39	43	52	61	69	87
3	07	13	26	39	46	52	59	65	78	91	1.04	1.30
4	09	17	35	52	61	69	78	87	1.04	1.21	1.39	1.73
5	11	22	43	65	76	87	98	1.08	1.30	1.52	1.73	2.17
6	13	26	52	78	91	1.04	1.17	1.30	1.56	1.82	2.08	2.60
7	15	30	61	91	1.06	1.21	1.37	1.52	1.82	2.12	2.43	3.03
8	17	35	69	1.04	1.21	1.39	1.56	1.73	2.08	2.43	2.77	3.47
9	20	39	78	1.17	1.37	1.56	1.76	1.95	2.34	2.73	3.12	3.90
10	22	43	87	1.30	1.52	1.73	1.95	2.17	2.60	3.03	3.47	4.33
11	24	48	95	1.43	1.67	1.91	2.15	2.38	2.86	3.34	3.81	4.77
Days.												
1	0	0	0	0	01	01	01	01	01	01	01	01
2	0	0	01	01	01	01	01	01	02	02	02	03
3	0	0	01	01	02	02	02	02	03	03	03	04
4	0	01	01	02	02	02	03	03	03	04	05	06
5	0	01	01	02	03	03	03	04	04	05	06	07
6	0	01	02	03	03	03	04	04	05	06	07	09
7	01	01	02	03	04	04	05	05	06	07	08	10
8	01	01	02	03	04	05	05	06	07	08	09	12
9	01	01	03	04	05	05	06	07	08	09	10	13
10	01	01	03	04	05	06	07	07	09	10	12	14
11	01	02	03	05	06	06	07	08	10	11	13	16
12	01	02	03	05	06	07	08	09	10	12	14	17
13	01	02	04	06	07	08	08	09	11	13	15	19
14	01	02	04	06	07	08	09	10	12	14	16	20
15	01	02	04	07	08	09	10	11	13	15	17	22
16	01	02	05	07	08	09	10	12	14	16	18	23
17	01	02	05	07	09	10	11	12	15	17	20	25
18	01	03	05	08	09	10	12	13	16	18	21	26
19	01	03	05	08	10	11	12	14	16	19	22	27
20	01	03	06	09	10	12	13	14	17	20	23	29
21	02	03	06	09	11	12	14	15	18	21	24	30
22	02	03	06	10	11	13	14	16	19	22	25	32
23	02	03	07	10	12	13	15	17	20	23	27	33
24	02	03	07	10	12	14	16	17	21	24	28	35
25	02	04	07	11	13	14	16	18	22	25	29	36
26	02	04	08	11	13	15	17	19	23	26	30	38
27	02	04	08	12	14	16	18	20	23	27	31	39
28	02	04	08	12	14	16	18	20	24	28	32	40
29	02	04	08	13	15	17	19	21	25	29	34	42
30	02	04	09	13	15	17	20	22	26	30	35	43
33	02	05	10	14	17	19	21	24	29	33	38	48
63	05	09	18	27	32	36	41	46	55	64	73	91
93	07	13	27	40	47	54	60	67	81	94	1.07	1.34

53 Dollars.

Years.	½ per ct.	1 per ct.	2 per ct.	3 per ct.	3½ per ct.	4 per ct.	4½ per ct.	5 per ct.	6 per ct.	7 per ct.	8 per ct.	10 per ct.
1	27	53	1.06	1.59	1.86	2.12	2.39	2.65	3.18	3.71	4.24	5.30
2	53	1.06	2.12	3.18	3.71	4.24	4.77	5.30	6.36	7.42	8.48	10.60
3	80	1.59	3.18	4.77	5.57	6.36	7.16	7.95	9.54	11.13	12.72	15.90
4	1.06	2.12	4.24	6.36	7.42	8.48	9.54	10.60	12.72	14.84	16.96	21.20
5	1.33	2.65	5.30	7.95	9.28	10.60	11.93	13.25	15.90	18.55	21.20	26.50
Months.												
1	02	04	09	13	15	18	20	22	27	31	35	44
2	04	09	18	27	31	35	40	44	53	62	71	88
3	07	13	27	40	46	53	60	66	80	93	1.06	1.33
4	09	18	35	53	62	71	80	88	1.06	1.24	1.41	1.77
5	11	22	44	66	77	88	99	1.10	1.33	1.55	1.77	2.21
6	13	27	53	80	93	1.06	1.19	1.33	1.59	1.86	2.12	2.65
7	15	31	62	93	1.08	1.24	1.39	1.55	1.86	2.16	2.47	3.09
8	18	35	71	1.06	1.24	1.41	1.59	1.77	2.12	2.47	2.83	3.53
9	20	40	80	1.19	1.39	1.59	1.79	1.99	2.39	2.78	3.18	3.98
10	22	44	88	1.33	1.55	1.77	1.99	2.21	2.65	3.09	3.53	4.42
11	24	49	97	1.46	1.70	1.94	2.19	2.43	2.92	3.40	3.89	4.86
Days.												
1	0	0	0	0	01	01	01	01	01	01	01	01
2	0	0	01	01	01	01	01	01	02	02	02	03
3	0	0	01	01	02	02	02	02	03	03	04	04
4	0	01	01	02	02	02	03	03	04	04	05	06
5	0	01	01	02	03	03	03	04	04	05	06	07
6	0	01	02	03	03	04	04	04	05	06	07	09
7	01	01	02	03	04	04	05	05	06	07	08	10
8	01	01	02	04	04	05	05	06	07	08	09	12
9	01	01	03	04	05	05	06	07	08	09	11	13
10	01	01	03	04	05	06	07	07	09	10	12	15
11	01	02	03	05	06	06	07	08	10	11	13	16
12	01	02	04	05	06	07	08	09	11	12	14	18
13	01	02	04	06	07	08	09	10	11	13	15	19
14	01	02	04	06	07	08	09	10	12	14	16	21
15	01	02	04.	07	08	09	10	11	13	15	18	22
16	01	02	05	07	08	09	11	12	14	16	19	24
17	01	03	05	08	09	10	11	13	15	18	20	25
18	01	03	05	08	09	11	12	13	16	19	21	27
19	01	03	06	08	10	11	13	14	17	20	22	28
20	01	03	06	09	10	12	13	15	18	21	24	29
21	02	03	06	09	11	12	14	15	19	22	25	31
22	02	03	06	10	11	13	15	16	19	23	26	32
23	02	03	07	10	12	14	15	17	20	24	27	34
24	02	04	07	11	12	14	16	18	21	25	28	35
25	02	04	07	11	13	15	17	18	22	26	29	37
26	02	04	08	11	13	15	17	19	23	27	31	38
27	02	04	08	12	14	16	18	20	24	28	32	40
28	02	04	08	12	14	16	19	21	25	29	33	41
29	02	04	09	13	15	17	19	21	26	30	34	43
30	02	04	09	13	15	18	20	22	27	31	35	44
33	02	05	10	15	17	19	22	24	29	34	39	49
63	05	09	19	28	32	37	42	46	56	65	74	93
93	07	14	27	41	48	55	62	68	82	96	1.10	1.37

54 Dollars.

	½ per ct.	1 per ct.	2 per ct.	3 per ct.	3½ per ct.	4 per ct.	4½ per ct.	5 per ct.	6 per ct.	7 per ct.	8 per ct.	10 per ct.
Years.												
1	27	54	1.08	1.62	1.89	2.16	2.43	2.70	3.24	3.78	4.32	5.40
2	54	1.08	2.16	3.24	3.78	4.32	4.86	5.40	6.48	7.56	8.64	10.80
3	81	1.62	3.24	4.86	5.67	6.48	7.29	8.10	9.72	11.34	12.96	16.20
4	1.08	2.16	4.32	6.48	7.56	8.64	9.72	10.80	12.96	15.12	17.28	21.60
5	1.35	2.70	5.40	8.10	9.45	10.80	12.15	13.50	16.20	18.90	21.60	27.00
Months.												
1	02	05	09	14	16	18	20	23	27	32	36	45
2	05	09	18	27	32	36	41	45	54	63	72	90
3	07	14	27	41	47	55	61	68	81	95	1.08	1.35
4	09	18	36	54	63	72	81	90	1.08	1.26	1.44	1.80
5	11	23	45	68	79	90	1.01	1.13	1.35	1.58	1.80	2.25
6	14	27	54	81	95	1.08	1.22	1.35	1.62	1.89	2.16	2.70
7	16	32	63	95	1.10	1.26	1.42	1.58	1.89	2.21	2.52	3.15
8	18	36	72	1.08	1.26	1.44	1.62	1.80	2.16	2.52	2.88	3.60
9	20	41	81	1.22	1.42	1.62	1.82	2.03	2.43	2.84	3.24	4.05
10	23	45	90	1.35	1.58	1.80	2.03	2.25	2.70	3.15	3.60	4.50
11	25	50	99	1.49	1.73	1.98	2.23	2.48	2.97	3.47	3.96	4.95
Days.												
1	0	0	0	0	01	01	01	01	01	01	01	02
2	0	0	01	01	01	01	01	02	02	02	02	03
3	0	0	01	01	02	02	02	02	03	03	04	05
4	0	01	01	02	02	02	03	03	04	04	05	06
5	0	01	02	02	03	03	03	04	05	05	06	08
6	0	01	02	03	03	04	04	05	05	06	07	09
7	01	01	02	03	04	04	05	05	06	07	08	11
8	01	01	02	04	04	05	05	06	07	08	10	12
9	01	01	03	04	05	05	06	07	08	09	11	14
10	01	02	03	05	05	06	07	08	09	11	12	15
11	01	02	03	05	06	07	07	08	10	12	13	17
12	01	02	04	05	06	07	08	09	11	13	14	18
13	01	02	04	06	07	08	09	10	12	14	16	20
14	01	02	04	06	07	08	09	11	13	15	17	21
15	01	02	05	07	08	09	10	11	14	16	18	23
16	01	02	05	07	08	10	11	12	14	17	19	24
17	01	03	05	08	09	10	11	13	15	18	20	26
18	01	03	05	08	09	11	12	14	16	19	22	27
19	01	03	06	09	10	11	13	14	17	20	23	29
20	02	03	06	09	11	12	14	15	18	21	24	30
21	02	03	06	09	11	13	14	16	19	22	25	32
22	02	03	07	10	12	13	15	17	20	23	26	33
23	02	03	07	10	12	14	16	17	21	24	28	35
24	02	04	07	11	13	14	16	18	22	25	29	36
25	02	04	08	11	13	15	17	19	23	26	30	38
26	02	04	08	12	14	16	18	20	23	27	31	39
27	02	04	08	12	14	16	18	20	24	28	32	41
28	02	04	08	13	15	17	19	21	25	29	34	42
29	02	04	09	13	15	17	20	22	26	30	35	44
30	02	05	09	14	16	18	20	23	27	32	36	45
33	02	05	10	15	17	20	22	25	30	35	40	50
63	05	10	19	29	33	38	43	48	57	67	76	96
93	07	14	28	42	49	56	63	70	84	98	1.12	1.40

55 Dollars.

Years.	½ per ct.	1 per ct.	2 per ct.	3 per ct.	3½ per ct.	4 per ct.	4½ per ct.	5 per ct.	6 per ct.	7 per ct.	8 per ct.	10 per ct.
1	28	55	1.10	1.65	1.93	2.20	2.48	2.75	3.30	3.85	4.40	5.50
2	55	1.10	2.20	3.30	3.85	4.40	4.75	5.50	6.60	7.70	8.80	11.00
3	83	1.65	3.30	4.95	5.78	6.60	7.43	8.25	9.90	11.55	13.20	16.50
4	1.10	2.20	4.40	6.60	7.70	8.80	9.90	11.00	13.20	15.40	17.60	22.00
5	1.38	2.75	5.50	8.25	9.63	11.00	12.38	13.75	16.50	19.25	22.00	27.50
Months.												
1	02	05	09	14	16	18	21	23	28	32	37	46
2	05	09	18	28	32	37	41	46	55	64	73	92
3	07	14	28	41	48	55	62	69	83	96	1.10	1.38
4	09	18	37	55	64	73	83	92	1.10	1.28	1.47	1.83
5	11	23	46	69	80	92	1.03	1.15	1.38	1.60	1.83	2.29
6	14	28	55	83	96	1.10	1.24	1.38	1.65	1.93	2.20	2.75
7	16	32	64	96	1.12	1.28	1.44	1.60	1.93	2.25	2.57	3.21
8	18	37	73	1.10	1.28	1.47	1.65	1.83	2.20	2.57	2.93	3.67
9	21	41	83	1.24	1.44	1.65	1.86	2.06	2.48	2.89	3.30	4.13
10	23	46	92	1.38	1.60	1.83	2.06	2.29	2.75	3.21	3.67	4.58
11	25	50	1.01	1.51	1.76	2.02	2.27	2.52	3.03	3.53	4.03	5.04
Days.												
1	0	0	0	0	01	01	01	01	01	01	01	02
2	0	0	01	01	01	01	01	02	02	02	02	03
3	0	0	01	01	02	02	02	02	03	03	04	05
4	0	01	01	02	02	02	03	03	04	04	05	06
5	0	01	02	02	03	03	03	04	05	05	06	08
6	0	01	02	03	03	04	04	05	06	06	07	09
7	01	01	02	03	04	04	05	05	06	07	09	11
8	01	01	02	04	04	05	06	06	07	09	10	12
9	01	01	03	04	05	06	06	07	08	10	11	14
10	01	02	03	05	05	06	07	08	09	11	12	15
11	01	02	03	05	06	07	08	08	10	12	13	17
12	01	02	04	06	06	07	08	09	11	13	15	18
13	01	02	04	06	07	08	09	10	12	14	16	20
14	01	02	04	06	07	09	10	11	13	15	17	21
15	01	02	05	07	08	09	10	11	14	16	18	23
16	01	02	05	07	09	10	11	12	15	17	20	24
17	01	03	05	08	09	10	12	13	16	18	21	26
18	01	03	06	08	10	11	12	14	17	19	22	28
19	01	03	06	09	10	12	13	15	17	20	23	29
20	02	03	06	09	11	12	14	15	18	21	24	31
21	02	03	06	10	11	13	14	16	19	22	26	32
22	02	03	07	10	12	13	15	17	20	24	27	34
23	02	04	07	11	12	14	16	18	21	25	28	35
24	02	04	07	11	13	15	17	18	22	26	29	37
25	02	04	08	11	13	15	17	19	23	27	31	38
26	02	04	08	12	14	16	18	20	24	28	32	40
27	02	04	08	12	14	17	19	21	25	29	33	41
28	02	04	09	13	15	17	19	21	26	30	34	43
29	02	04	09	13	16	18	20	22	27	31	35	44
30	02	05	09	14	16	18	21	23	28	32	37	46
33	03	05	10	15	18	20	23	25	30	35	40	50
63	05	10	19	29	34	39	43	48	58	67	77	96
93	07	14	28	43	50	57	64	71	85	99	1.14	1.42

56 Dollars.

Years.	½ per ct.	1 per ct.	2 per ct.	3 per ct.	3½ per ct.	4 per ct.	4½ per ct.	5 per ct.	6 per ct.	7 per ct.	8 per ct.	10 per ct.
1	28	56	1.12	1.68	1.96	2.24	2.52	2.80	3.36	3.92	4.48	5.60
2	56	1.12	2.24	3.36	3.92	4.48	5.04	5.60	6.72	7.84	8.96	11.20
3	84	1.68	3.36	5.04	5.88	6.72	7.56	8.40	10.08	11.76	13.44	16.80
4	1.12	2.24	4.48	6.72	7.84	8.96	10.08	11.20	13.44	15.68	17.92	22.40
5	1.40	2.80	5.60	8.40	9.80	11.20	12.60	14.00	16.80	19.60	22.40	28.00
Months.												
1	02	05	09	14	16	19	21	23	28	33	37	47
2	05	09	19	28	33	37	42	47	56	65	75	93
3	07	14	28	42	49	56	63	70	84	98	1.12	1.40
4	09	19	37	56	65	75	84	93	1.12	1.31	1.49	1.87
5	12	23	47	70	82	93	1.05	1.17	1.40	1.63	1.87	2.33
6	14	28	56	84	98	1.12	1.26	1.40	1.68	1.96	2.24	2.80
7	16	33	65	98	1.14	1.31	1.47	1.63	1.96	2.29	2.61	3.27
8	19	37	75	1.12	1.31	1.49	1.68	1.87	2.24	2.61	2.99	3.73
9	21	42	84	1.26	1.47	1.68	1.89	2.10	2.52	2.94	3.36	4.20
10	23	47	93	1.40	1.63	1.87	2.10	2.33	2.80	3.27	3.73	4.67
11	26	51	1.03	1.54	1.80	2.05	2.31	2.57	3.08	3.59	4.11	5.13
Days.												
1	0	0	0	0	01	01	01	01	01	01	01	02
2	0	0	01	01	01	01	01	02	02	02	02	03
3	0	0	01	01	02	02	02	02	03	03	04	05
4	0	01	01	02	02	02	03	03	04	04	05	06
5	0	01	02	02	03	03	04	04	05	05	06	08
6	0	01	02	03	03	04	04	05	06	07	07	09
7	01	01	02	03	04	04	05	05	07	08	09	11
8	01	01	02	04	04	05	06	06	07	09	10	12
9	01	01	03	04	05	06	06	07	08	10	11	14
10	01	02	03	05	05	06	07	08	09	11	12	16
11	01	02	03	05	06	07	08	09	10	12	14	17
12	01	02	04	06	07	07	08	09	11	13	15	19
13	01	02	04	06	07	08	09	10	12	14	16	20
14	01	02	04	07	08	09	10	11	13	15	17	22
15	01	02	05	07	08	09	11	12	14	16	19	23
16	01	02	05	07	09	10	11	12	15	17	20	25
17	01	03	05	08	09	11	12	13	16	19	21	26
18	01	03	06	08	10	11	13	14	17	20	22	28
19	01	03	06	09	10	12	13	15	18	21	24	30
20	02	03	06	09	11	12	14	16	19	22	25	31
21	02	03	07	10	11	13	15	16	20	23	26	33
22	02	03	07	10	12	14	15	17	21	24	27	34
23	02	04	07	11	13	14	16	18	21	25	29	36
24	02	04	07	11	13	15	17	19	22	26	30	37
25	02	04	08	12	14	16	18	19	23	27	31	39
26	02	04	08	12	14	16	18	20	24	28	32	40
27	02	04	08	13	15	17	19	21	25	29	34	42
28	02	04	09	13	15	17	20	22	26	30	35	44
29	02	05	09	14	16	18	20	23	27	32	36	45
30	02	05	09	14	16	19	21	23	28	33	37	47
33	03	05	10	15	18	21	23	26	31	36	41	51
63	05	10	20	29	34	39	44	49	59	69	78	98
93	07	14	29	43	51	58	65	72	87	1.01	1.16	1.45

57 Dollars.

Years.	½ per ct.	1 per ct.	2 per ct.	3 per ct.	3½ per ct.	4 per ct.	4½ per ct.	5 per ct.	6 per ct.	7 per ct.	8 per ct.	10 per ct.
1	29	57	1.14	1.71	2.00	2.28	2.57	2.85	3.42	3.99	4.56	5.70
2	57	1.14	2.28	3.42	3.99	4.56	5.13	5.70	6.84	7.98	9.12	11.40
3	85	1.71	3.42	5.13	5.99	6.84	7.70	8.55	10.26	11.97	13.68	17.10
4	1.14	2.28	4.56	6.84	7.98	9.12	10.30	11.40	13.68	15.96	18.24	22.80
5	1.43	2.85	5.70	8.55	9.98	11.40	12.83	14.25	17.10	19.95	22.80	28.50
Months.												
1	02	05	10	14	17	19	21	24	29	33	38	48
2	05	10	19	29	33	38	43	48	57	66	76	95
3	07	14	29	43	50	57	64	71	86	1.00	1.14	1.43
4	10	19	38	57	67	76	86	95	1.14	1.33	1.52	1.90
5	12	24	48	71	83	95	1.07	1.19	1.43	1.66	1.90	2.38
6	14	29	57	86	1.00	1.14	1.28	1.43	1.71	2.00	2.28	2.85
7	17	33	67	1.00	1.16	1.33	1.50	1.66	2.00	2.33	2.66	3.33
8	19	38	76	1.14	1.33	1.52	1.71	1.90	2.28	2.66	3.04	3.80
9	21	43	86	1.28	1.49	1.71	1.92	2.14	2.57	2.99	3.42	4.28
10	24	48	95	1.43	1.66	1.90	2.14	2.38	2.85	3.33	3.80	4.75
11	26	52	1.05	1.57	1.83	2.09	2.35	2.61	3.14	3.66	4.18	5.23
Days.												
1	0	0	0	0	01	01	01	01	01	01	01	02
2	0	0	01	01	01	01	01	02	02	02	03	03
3	0	0	01	01	02	02	02	02	03	03	04	05
4	0	01	01	02	02	03	03	03	04	04	05	06
5	0	01	02	02	03	03	04	04	05	06	06	08
6	0	01	02	03	03	04	04	05	06	07	08	10
7	01	01	02	03	04	04	05	06	07	08	09	11
8	01	01	03	04	04	05	06	06	08	09	10	13
9	01	01	03	04	05	06	06	07	09	10	11	14
10	01	02	03	05	06	06	07	08	10	11	13	16
11	01	02	03	05	06	07	08	09	10	12	14	17
12	01	02	04	06	07	08	09	10	11	13	15	19
13	01	02	04	06	07	08	09	10	12	14	16	21
14	01	02	04	07	08	09	10	11	13	16	18	22
15	01	02	05	07	08	10	11	12	14	17	19	24
16	01	03	05	08	09	10	11	13	15	18	20	25
17	01	03	05	08	09	11	12	13	16	19	22	27
18	01	03	06	09	10	11	13	14	17	20	23	29
19	02	03	06	09	11	12	14	15	18	21	24	30
20	02	03	06	10	11	13	14	16	19	22	25	32
21	02	03	07	10	12	13	15	17	20	23	27	33
22	02	03	07	10	12	14	16	17	21	24	28	35
23	02	04	07	11	13	15	16	18	22	25	29	36
24	02	04	08	11	13	15	17	19	23	27	30	38
25	02	04	08	12	14	16	18	20	24	28	32	40
26	02	04	08	12	14	16	19	21	25	29	33	41
27	02	04	09	13	15	17	19	21	26	30	34	43
28	02	04	09	13	16	18	20	22	27	31	35	44
29	02	05	09	14	16	18	21	23	28	32	37	46
30	02	05	10	14	17	19	21	24	29	33	38	48
33	03	05	10	16	18	21	24	26	31	37	42	52
63	05	10	20	30	35	40	45	50	60	70	80	1.00
93	07	15	29	44	52	59	66	74	88	1.03	1.18	1.47

58 Dollars.

Years.	½ per ct.	1 per ct.	2 per ct.	3 per ct.	3½ per ct.	4 per ct.	4½ per ct.	5 per ct.	6 per ct.	7 per ct.	8 per ct.	10 per ct.
1	29	58	1.16	1.74	2.03	2.32	2.61	2.90	3.48	4.06	4.64	5.80
2	58	1.16	2.32	3.48	4.06	4.64	5.22	5.80	6.96	8.12	9.28	11.60
3	87	1.74	3.48	5.22	6.09	6.96	7.83	8.70	10.44	12.18	13.92	17.40
4	1.16	2.32	4.64	6.96	8.12	9.28	10.44	11.60	13.92	16.24	18.56	23.20
5	1.45	2.90	5.80	8.70	10.15	11.60	13.05	14.50	17.40	20.30	23.20	29.00
Months.												
1	02	05	10	15	17	19	22	24	29	34	39	48
2	05	10	19	29	34	39	44	48	58	68	77	97
3	07	15	29	44	51	58	65	73	87	1.02	1.16	1.45
4	10	19	39	58	68	77	87	97	1.16	1.35	1.55	1.93
5	12	24	48	73	85	97	1.09	1.21	1.45	1.69	1.93	2.42
6	15	29	58	87	1.02	1.16	1.31	1.45	1.74	2.03	2.32	2.90
7	17	34	68	1.02	1.18	1.35	1.52	1.69	2.03	2.37	2.71	3.38
8	19	39	77	1.16	1.35	1.55	1.74	1.93	2.32	2.71	3.09	3.87
9	22	44	87	1.31	1.52	1.74	1.96	2.18	2.61	3.05	3.48	4.35
10	24	48	97	1.45	1.69	1.93	2.18	2.42	2.90	3.38	3.87	4.83
11	27	53	1.06	1.60	1.86	2.13	2.39	2.66	3.19	3.72	4.25	5.32
Days.												
1	0	0	0	0	01	01	01	01	01	01	01	02
2	0	0	01	01	01	01	01	02	02	02	03	03
3	0	0	01	01	02	02	02	02	03	03	04	05
4	0	01	01	02	02	03	03	03	04	05	05	06
5	0	01	02	02	03	03	04	04	05	06	06	08
6	0	01	02	03	03	04	04	05	06	07	08	10
7	01	01	02	03	04	05	05	06	07	08	09	11
8	01	01	03	04	05	05	06	06	08	09	10	13
9	01	01	03	04	05	06	07	07	09	10	12	15
10	01	02	03	05	06	06	07	08	10	11	13	16
11	01	02	04	05	06	07	08	09	11	12	14	18
12	01	02	04	06	07	08	09	10	12	14	15	19
13	01	02	04	06	07	08	09	10	13	15	17	21
14	01	02	05	07	08	09	10	11	14	16	18	23
15	01	02	05	07	08	10	11	12	15	17	19	24
16	01	03	05	08	09	10	12	13	15	18	21	26
17	01	03	05	08	10	11	12	14	16	19	22	27
18	01	03	06	09	10	12	13	15	17	20	23	29
19	02	03	06	09	11	12	14	15	18	21	24	31
20	02	03	06	10	11	13	15	16	19	23	26	32
21	02	03	07	10	12	14	15	17	20	24	27	34
22	02	04	07	11	12	14	16	18	21	25	28	35
23	02	04	07	11	13	15	17	19	22	26	30	37
24	02	04	08	12	14	15	17	19	23	27	31	39
25	02	04	08	12	14	16	18	20	24	28	32	40
26	02	04	08	13	15	17	19	21	25	29	34	42
27	02	04	09	13	15	17	20	22	26	30	35	44
28	02	05	09	14	16	18	20	23	27	32	36	45
29	02	05	09	14	16	19	21	23	28	33	37	47
30	02	05	10	15	17	19	22	24	29	34	39	48
33	03	05	10	16	18	21	24	26	31	37	42	52
63	05	10	20	30	36	41	46	51	61	71	81	1.02
93	07	15	30	45	52	60	67	75	90	1.05	1.20	1.50

59 Dollars.

Years.	½ per ct.	1 per ct.	2 per ct.	3 per ct.	3½ per ct.	4 per ct.	4½ per ct.	5 per ct.	6 per ct.	7 per ct.	8 per ct.	10 per ct.
1	30	59	1.18	1.77	2.07	2.36	2.66	2.95	3.54	4.13	4.72	5.90
2	59	1.18	2.36	3.54	4.13	4.72	5.31	5.90	7.08	8.26	9.44	11.80
3	89	1.77	3.54	5.31	6.20	7.08	7.97	8.85	10.62	12.39	14.16	17.70
4	1.18	2.36	4.72	7.08	8.26	9.44	10.62	11.80	14.16	16.52	18.88	23.60
5	1.48	1.45	5.90	8.85	10.33	11.80	13.28	14.75	17.70	20.65	23.60	29.50
Months.												
1	02	05	10	15	17	20	22	25	30	34	39	49
2	05	10	20	30	34	39	44	49	59	69	79	98
3	07	15	30	44	52	59	66	74	89	1.03	1.18	1.48
4	10	20	39	59	69	79	89	98	1.18	1.38	1.57	1.97
5	12	25	49	74	86	98	1.11	1.23	1.48	1.72	1.97	2.46
6	15	30	59	89	1.03	1.18	1.33	1.48	1.77	2.07	2.36	2.95
7	17	34	69	1.03	1.20	1.38	1.55	1.72	2.07	2.41	2.75	3.44
8	20	39	79	1.18	1.38	1.57	1.77	1.97	2.36	2.75	3.15	3.93
9	22	44	89	1.33	1.55	1.77	1.99	2.21	2.66	3.10	3.54	4.43
10	25	49	98	1.48	1.72	1.97	2.21	2.46	2.95	3.44	3.93	4.92
11	27	54	1.08	1.62	1.89	2.16	2.43	2.70	3.25	3.79	4.33	5.41
Days.												
1	0	0	0	0	01	01	01	01	01	01	01	02
2	0	0	01	01	01	01	01	02	02	02	03	03
3	0	0	01	01	02	02	02	02	03	03	04	05
4	0	01	01	02	02	03	03	03	04	05	05	07
5	0	01	02	02	03	03	04	04	05	06	07	08
6	0	01	02	03	03	04	04	05	06	07	08	10
7	01	01	02	03	04	05	05	06	07	08	09	11
8	01	01	03	04	05	05	06	07	08	09	10	13
9	01	01	03	04	05	06	07	07	09	10	12	15
10	01	02	03	05	06	07	07	08	10	11	13	16
11	01	02	04	05	06	07	08	09	11	13	14	18
12	01	02	04	06	07	08	09	10	12	14	16	20
13	01	02	04	06	07	09	10	11	13	15	17	21
14	01	02	05	07	08	09	10	11	14	16	18	23
15	01	02	05	07	09	10	11	12	15	17	20	24
16	01	03	05	08	09	10	12	13	16	18	21	26
17	01	03	06	08	10	11	13	14	17	20	22	28
18	01	03	06	09	10	12	13	15	18	21	24	30
19	02	03	06	09	11	12	14	16	19	22	25	31
20	02	03	07	10	11	13	15	16	20	23	26	33
21	02	03	07	10	12	14	15	17	21	24	28	34
22	02	04	07	11	13	14	16	18	22	25	29	36
23	02	04	08	11	13	15	17	19	23	26	30	38
24	02	04	08	12	14	16	18	20	24	28	31	39
25	02	04	08	12	14	16	18	20	25	29	33	41
26	02	04	09	13	15	17	19	21	26	30	34	43
27	02	04	09	13	15	18	20	22	27	31	35	44
28	02	05	09	14	16	18	21	23	28	32	37	46
29	02	05	10	14	17	19	21	24	29	33	38	48
30	02	05	10	15	17	20	22	25	30	34	39	49
33	03	05	11	16	19	22	24	27	32	38	43	54
63	05	10	21	31	36	41	46	52	62	72	83	1.03
93	08	15	30	46	53	61	69	76	91	1.07	1.22	1.52

60 Dollars.

	½ per ct.	1 per ct.	2 per ct.	3 per ct.	3½ per ct.	4 per ct.	4½ per ct.	5 per ct.	6 per ct.	7 per ct.	8 per ct.	10 per ct.
Years.												
1	30	60	1.20	1.80	2.10	2.40	2.70	3.00	3.60	4.20	4.80	6.00
2	60	1.20	2.40	3.60	4.20	4.80	5.40	6.00	7.20	8.40	9.60	12.00
3	90	1.80	3.60	5.40	6.30	7.20	8.10	9.00	10.80	12.60	14.40	18.00
4	1.20	2.40	4.80	7.20	8.40	9.60	10.80	12.00	14.40	16.80	19.20	24.00
5	1.50	3.00	6.00	9.00	10.50	12.00	13.50	15.00	18.00	21.00	24.00	30.00
Months.												
1	03	05	10	15	18	20	23	25	30	35	40	50
2	05	10	20	30	35	40	45	50	60	70	80	1.00
3	08	15	30	45	53	60	68	75	90	1.05	1.20	1.50
4	10	20	40	60	70	80	90	1.00	1.20	1.40	1.60	2.00
5	13	25	50	75	88	1.00	1.13	1.25	1.50	1.75	2.00	2.50
6	15	30	60	90	1.05	1.20	1.35	1.50	1.80	2.10	2.40	3.00
7	18	35	70	1.05	1.23	1.40	1.58	1.75	2.10	2.45	2.80	3.50
8	20	40	80	1.20	1.40	1.60	1.80	2.00	2.40	2.80	3.20	4.00
9	23	45	90	1.35	1.58	1.80	2.03	2.25	2.70	3.15	3.60	4.50
10	25	50	1.00	1.50	1.75	2.00	2.25	2.50	3.00	3.50	4.00	5.00
11	28	55	1.10	1.65	1.98	2.20	2.48	2.75	3.30	3.85	4.40	5.50
Days.												
1	0	0	0	01	01	01	01	01	01	01	01	02
2	0	0	01	01	01	01	01	02	02	02	03	03
3	0	01	01	02	02	02	02	03	03	04	04	05
4	0	01	01	02	02	03	03	03	04	05	05	07
5	0	01	02	03	03	03	04	04	05	06	07	08
6	01	01	02	03	04	04	05	05	06	07	08	10
7	01	01	02	04	04	05	05	06	07	08	09	12
8	01	01	03	04	05	05	06	07	08	09	11	13
9	01	02	03	05	05	06	07	08	09	11	12	15
10	01	02	03	05	06	07	08	08	10	12	13	17
11	01	02	04	06	06	07	08	09	11	13	15	18
12	01	02	04	06	07	08	09	10	12	14	16	20
13	01	02	04	07	08	09	10	11	13	15	17	22
14	01	02	05	07	08	09	11	12	14	16	19	23
15	01	03	05	08	09	10	11	13	15	18	20	25
16	01	03	05	08	09	11	12	13	16	19	21	27
17	01	03	06	09	10	11	13	14	17	20	23	28
18	02	03	06	09	11	12	14	15	18	21	24	30
19	02	03	06	10	11	13	14	16	19	22	25	32
20	02	03	07	10	12	13	15	17	20	23	27	33
21	02	04	07	11	12	14	16	18	21	25	28	35
22	02	04	07	11	13	15	17	18	22	26	29	37
23	02	04	08	12	13	15	17	19	23	27	31	38
24	02	04	08	12	14	16	18	20	24	28	32	40
25	02	04	08	13	15	17	19	21	25	29	33	42
26	02	04	09	13	15	17	20	22	26	30	35	43
27	02	05	09	14	16	18	20	23	27	32	36	45
28	02	05	09	14	16	19	21	23	28	33	37	47
29	02	05	10	15	17	19	22	24	29	34	39	48
30	03	05	10	15	18	20	23	25	30	35	40	50
33	03	06	11	17	19	22	25	28	33	39	44	55
63	05	11	21	32	37	42	47	53	63	74	84	1.05
93	08	16	31	47	54	62	70	78	93	1.09	1.24	1.55

61 Dollars.

Years.	½ per ct.	1 per ct.	2 per ct.	3 per ct.	3½ per ct.	4 per ct.	4½ per ct.	5 per ct.	6 per ct.	7 per ct.	8 per ct.	10 per ct.
1	31	61	1.22	1.83	2.14	2.44	2.75	3.05	3.66	4.27	4.83	6.10
2	61	1.22	2.44	3.66	4.27	4.88	5.49	6.10	7.32	8.54	9.76	12.20
3	92	1.83	3.66	5.49	6.41	7.32	8.24	9.15	10.98	12.81	14.64	18.30
4	1.22	2.44	4.88	7.32	8.54	9.76	10.98	12.20	14.64	17.08	19.52	24.40
5	1.53	3.05	6.10	9.15	10.68	12.20	13.73	15.25	18.30	21.35	24.40	30.50
Months.												
1	03	05	10	15	18	20	23	25	31	36	41	51
2	05	10	20	31	36	41	46	51	61	71	81	1.02
3	08	15	31	46	53	61	69	76	92	1.07	1.22	1.53
4	10	20	41	61	71	81	92	1.02	1.22	1.42	1.63	2.03
5	13	25	51	76	89	1.02	1.14	1.27	1.53	1.78	2.03	2.54
6	15	31	61	92	1.07	1.22	1.37	1.53	1.83	2.14	2.44	3.05
7	18	36	71	1.06	1.25	1.42	1.60	1.78	2.14	2.49	2.85	3.56
8	20	41	81	1.22	1.42	1.63	1.83	2.03	2.44	2.85	3.25	4.07
9	23	46	92	1.37	1.60	1.83	2.06	2.29	2.75	3.20	3.66	4.58
10	25	51	1.02	1.53	1.78	2.03	2.29	2.54	3.05	3.56	4.07	5.08
11	28	56	1.12	1.68	1.96	2.24	2.52	2.80	3.36	3.91	4.47	5.59
Days.												
1	0	0	0	01	01	01	01	01	01	01	01	02
2	0	0	01	01	01	01	02	02	02	02	03	03
3	0	01	01	02	02	02	02	03	03	04	04	05
4	0	01	01	02	02	03	03	03	04	05	05	07
5	0	01	02	03	03	03	04	04	05	06	07	08
6	01	01	02	03	04	04	05	05	06	07	08	10
7	01	01	02	04	04	05	05	06	07	08	09	12
8	01	01	03	04	05	05	06	07	08	09	11	14
9	01	02	03	05	05	06	07	08	09	11	12	15
10	01	02	03	05	06	07	08	08	10	12	14	17
11	01	02	04	06	07	07	08	09	11	13	15	19
12	01	02	04	06	07	08	09	10	12	14	16	20
13	01	02	04	07	08	09	10	11	13	15	18	22
14	01	02	05	07	08	09	11	12	14	17	19	24
15	01	03	05	08	09	10	11	13	15	18	20	25
16	01	03	05	08	09	11	12	14	16	19	22	27
17	01	03	06	09	10	12	13	14	17	20	23	29
18	02	03	06	09	11	12	14	15	18	21	24	31
19	02	03	06	10	11	13	14	16	19	23	26	32
20	02	03	07	10	12	14	15	17	20	24	27	34
21	02	04	07	11	12	14	16	18	21	25	28	36
22	02	04	07	11	13	15	17	19	22	26	30	37
23	02	04	08	12	14	16	18	19	23	27	31	39
24	02	04	08	12	14	16	18	20	24	28	33	41
25	02	04	08	13	15	17	19	21	25	30	34	42
26	02	04	09	13	15	18	20	22	26	31	35	44
27	02	05	09	14	16	18	21	23	27	32	37	46
28	02	05	09	14	17	19	21	24	28	33	38	47
29	02	05	10	15	17	20	22	25	29	34	39	49
30	03	05	10	15	18	20	23	25	31	36	41	51
33	03	06	11	17	20	22	25	28	34	39	45	56
63	05	11	21	32	37	43	48	53	64	75	85	1.07
93	08	16	32	47	55	63	71	79	95	1.10	1.26	1.58

62 Dollars

Years.	½ per ct.	1 per ct.	2 per ct.	3 per ct.	3½ per ct.	4 per ct.	4½ per ct.	5 per ct.	6 per ct.	7 per ct.	8 per ct.	10 per ct.
1	31	62	1.24	1.86	2.17	2.48	2.79	3.10	3.72	4.34	4.96	6.20
2	62	1.24	2.48	3.72	4.34	4.96	5.58	6.20	7.44	8.68	9.92	12.40
3	93	1.86	3.72	5.58	6.51	7.44	8.37	9.30	11.16	13.02	14.88	18.60
4	1.24	2.48	4.96	7.44	8.68	9.92	11.16	12.40	14.88	17.36	19.84	24.80
5	1.55	3.10	6.20	9.30	10.85	12.40	13.95	15.50	18.60	21.70	24.80	31.00
Months.												
1	03	05	10	16	18	21	23	26	31	36	41	52
2	05	10	21	31	36	41	47	52	62	72	83	1.03
3	08	16	31	47	54	62	70	78	93	1.09	1.24	1.55
4	10	21	41	62	72	83	93	1.03	1.24	1.45	1.65	2.07
5	13	26	52	78	90	1.03	1.16	1.29	1.55	1.81	2.07	2.58
6	16	31	62	93	1.09	1.24	1.40	1.55	1.86	2.17	2.48	3.10
7	18	36	72	1.09	1.27	1.45	1.63	1.81	2.17	2.53	2.89	3.62
8	21	41	83	1.24	1.45	1.65	1.86	2.07	2.48	2.89	3.31	4.13
9	23	47	93	1.40	1.63	1.86	2.09	2.33	2.79	3.26	3.72	4.65
10	26	52	1.03	1.55	1.81	2.07	2.33	2.58	3.10	3.62	4.13	5.17
11	28	57	1.14	1.71	1.99	2.27	2.56	2.84	3.41	3.98	4.55	5.68
Days.												
1	0	0	0	01	01	01	01	01	01	01	01	02
2	0	0	01	01	01	01	02	02	02	02	03	03
3	0	01	01	02	02	02	02	03	03	04	04	05
4	0	01	01	02	02	03	03	03	04	05	06	07
5	0	01	02	03	03	03	04	04	05	06	07	09
6	01	01	02	03	04	04	05	05	06	07	08	10
7	01	01	02	04	04	05	05	06	07	08	10	12
8	01	01	03	04	05	06	06	07	08	10	11	14
9	01	02	03	05	05	06	07	08	09	11	12	16
10	01	02	03	05	06	07	08	09	10	12	14	17
11	01	02	04	06	07	08	09	09	11	13	15	19
12	01	02	04	06	07	08	09	10	12	14	17	21
13	01	02	04	07	08	09	10	11	13	16	18	22
14	01	02	05	07	08	10	11	12	14	17	19	24
15	01	03	05	08	09	10	12	13	16	18	21	26
16	01	03	06	08	10	11	12	14	17	19	22	28
17	01	03	06	09	10	12	13	15	18	20	23	29
18	02	03	06	09	11	12	14	16	19	22	25	31
19	02	03	07	10	11	13	15	16	20	23	26	33
20	02	03	07	10	12	14	16	17	21	24	28	34
21	02	04	07	11	13	14	16	18	22	25	29	36
22	02	04	08	11	13	15	17	19	23	27	30	38
23	02	04	08	12	14	16	18	20	24	28	32	40
24	02	04	08	12	14	17	19	21	25	29	33	41
25	02	04	09	13	15	17	19	22	26	30	34	43
26	02	04	09	13	16	18	20	22	27	31	36	45
27	02	05	09	14	16	19	21	23	28	33	37	47
28	02	05	10	14	17	19	22	24	29	34	39	48
29	02	05	10	15	17	20	22	25	30	35	40	50
30	03	05	10	16	18	21	23	26	31	36	41	52
33	03	06	11	17	20	23	26	28	34	40	45	57
63	05	11	22	33	38	43	49	54	65	76	87	1.09
93	08	16	32	48	56	64	72	80	96	1.12	1.28	1.60

63 Dollars.

	½ per ct.	1 per ct.	2 per ct.	3 per ct.	3½ per ct.	4 per ct.	4½ per ct.	5 per ct.	6 per ct.	7 per ct.	8 per ct.	10 per ct.
Years.												
1	32	63	1.26	1.89	2.21	2.52	2.84	3.15	3.78	4.41	5.04	6.30
2	63	1.26	2.52	3.78	4.42	5.04	5.67	6.30	7.56	8.82	10.08	12.60
3	95	1.89	3.78	5.67	6.62	7.56	8.51	9.45	11.34	13.23	15.12	18.90
4	1.26	2.52	5.04	7.56	8.82	10.08	11.34	12.60	15.12	17.64	20.16	25.20
5	1.58	3.15	6.30	9.45	11.03	12.60	14.18	15.75	18.90	22.05	25.20	31.50
Months.												
1	03	05	11	16	18	21	24	26	32	37	42	53
2	05	11	21	32	37	42	47	53	63	74	84	1.05
3	08	16	32	47	55	63	71	79	95	1.10	1.26	1 58
4	11	21	42	63	74	84	95	1.05	1.26	1.47	1.68	2.10
5	13	26	53	79	92	1.05	1.18	1.31	1.58	1.84	2.10	2.63
6	16	32	63	95	1.10	1.26	1.42	1.58	1.89	2.21	2.52	3.15
7	18	37	74	1.10	1.29	1.47	1.65	1.84	2.21	2.57	2.94	3.68
8	21	42	84	1.26	1.47	1.68	1.89	2.10	2.52	2.94	3.36	4.20
9	24	47	95	1.42	1.65	1.89	2.12	2.36	2.84	3.31	3.78	4.73
10	26	53	1.05	1.58	1.84	2.10	2.36	2.63	3.15	3.68	4.20	5.25
11	29	58	1.16	1.73	2.02	2.31	2.60	2.89	3.47	4.04	4.62	5.78
Days.												
1	0	0	0	01	01	01	01	01	01	01	01	02
2	0	0	01	01	01	01	02	02	02	02	03	04
3	0	01	01	02	02	02	02	03	03	04	04	05
4	0	01	01	02	02	03	03	04	04	05	06	07
5	0	01	02	03	03	04	04	04	05	06	07	09
6	01	01	02	03	04	04	05	05	06	07	08	11
7	01	01	02	04	04	05	06	06	07	09	10	12
8	01	01	03	04	05	06	06	07	08	10	11	14
9	01	02	03	05	06	06	07	08	09	11	13	16
10	01	02	04	05	06	07	08	09	11	12	14	18
11	01	02	04	06	07	08	09	10	12	13	15	19
12	01	02	04	06	07	08	09	11	13	15	17	21
13	01	02	05	07	08	09	10	11	14	16	18	23
14	01	02	05	07	09	10	11	12	15	17	20	25
15	01	03	05	08	09	11	12	13	16	18	21	26
16	01	03	06	08	10	11	13	14	17	20	22	28
17	01	03	06	09	10	12	13	15	18	21	24	30
18	02	03	06	09	11	13	14	16	19	22	25	32
19	02	03	07	10	12	13	15	17	20	23	27	33
20	02	04	07	11	12	14	16	18	21	25	28	35
21	02	04	07	11	13	15	17	18	22	26	29	37
22	02	04	08	12	13	15	17	19	23	27	31	39
23	02	04	08	12	14	16	18	20	24	28	32	40
24	02	04	08	13	15	17	19	21	25	29	34	42
25	02	04	09	13	15	18	20	22	26	31	35	44
26	02	05	09	14	16	18	20	23	27	32	36	46
27	02	05	09	14	17	19	21	24	28	33	38	47
28	02	05	10	15	17	20	22	25	29	34	39	49
29	03	05	10	15	18	20	23	25	30	36	41	51
30	03	05	11	16	18	21	24	26	32	37	42	53
33	03	06	12	17	20	23	26	29	35	40	46	58
63	05	11	22	33	39	44	50	55	66	77	88	1.10
93	08	16	33	49	57	65	73	81	98	1.14	1.30	1.63

64 Dollars.

Years.	½ per ct.	1 per ct.	2 per ct.	3 per ct.	3½ per ct.	4 per ct.	4½ per ct.	5 per ct.	6 per ct.	7 per ct.	8 per ct.	10 per ct.
1	32	64	1.28	1.92	2.24	2.56	2.88	3.20	3.84	4.48	5.12	6.40
2	64	1.28	2.56	3.84	4.48	5.12	5.76	6.40	7.68	8.96	10.24	12.80
3	96	1.92	3.84	5.76	6.72	7.68	8.64	9.60	11.52	13.44	15.36	19.20
4	1.28	2.56	5.12	7.68	8.96	10.24	11.52	12.80	15.36	17.92	20.48	25.60
5	1.60	3.20	6.40	9.60	11.20	12.80	14.40	16.00	19.20	22.40	25.60	32.00
Months.												
1	03	05	11	16	19	21	24	27	32	37	43	53
2	05	11	21	32	37	43	48	53	64	75	85	1.07
3	08	16	32	48	56	64	72	80	96	1.12	1.28	1.60
4	11	21	43	64	75	85	96	1.07	1.28	1.49	1.71	2.13
5	13	27	53	80	93	1.07	1.20	1.33	1.60	1.87	2.13	2.67
6	16	32	64	96	1.12	1.28	1.44	1.60	1.92	2.24	2.56	3.20
7	19	37	75	1.12	1.31	1.49	1.68	1.87	2.24	2.61	2.99	3.73
8	21	43	85	1.28	1.49	1.71	1.92	2.13	2.56	2.99	3.41	4.27
9	24	48	96	1.44	1.68	1.92	2.16	2.40	2.88	3.36	3.84	4.80
10	27	53	1.07	1.60	1.87	2.13	2.40	2.67	3.20	3.73	4.27	5.33
11	29	59	1.17	1.76	2.05	2.35	2.64	2.93	3.52	4.11	4.69	5.87
Days.												
1	0	0	0	01	01	01	01	01	01	01	01	02
2	0	0	01	01	01	01	02	02	02	02	03	04
3	0	01	01	02	02	02	02	03	03	04	04	05
4	0	01	01	02	02	03	03	04	04	05	06	07
5	0	01	02	03	03	04	04	04	05	06	07	09
6	01	01	02	03	04	04	05	05	06	07	09	11
7	01	01	02	04	04	05	06	06	07	09	10	12
8	01	01	03	04	05	06	06	07	09	10	11	14
9	01	02	03	05	06	06	07	08	10	11	13	16
10	01	02	04	05	06	07	08	09	11	12	14	18
11	01	02	04	06	07	08	09	10	12	14	16	20
12	01	02	04	06	07	09	10	11	13	15	17	21
13	01	02	05	07	08	09	10	12	14	16	18	23
14	01	02	05	07	09	10	11	12	15	17	20	25
15	01	03	05	08	09	11	12	13	16	19	21	27
16	01	03	06	09	10	11	13	14	17	20	23	28
17	02	03	06	09	11	12	14	15	18	21	24	30
18	02	03	06	10	11	13	14	16	19	22	26	32
19	02	03	07	10	12	14	15	17	20	24	27	34
20	02	04	07	11	12	14	16	18	21	25	28	36
21	02	04	07	11	13	15	17	19	22	26	30	37
22	02	04	08	12	14	16	18	20	23	27	31	39
23	02	04	08	12	14	16	18	20	25	29	33	41
24	02	04	09	13	15	17	19	21	26	30	34	43
25	02	04	09	13	16	18	20	22	27	31	36	44
26	02	05	09	14	16	18	21	23	28	32	37	46
27	02	05	10	14	17	19	22	24	29	34	38	48
28	02	05	10	15	17	20	22	25	30	35	40	50
29	03	05	10	15	18	21	23	26	31	36	41	52
30	03	05	11	16	19	21	24	27	32	37	43	53
33	03	06	12	18	21	23	26	29	35	41	47	59
63	05	11	22	34	39	45	50	56	67	78	90	1.12
93	08	17	33	50	58	66	74	83	99	1.16	1.32	1.65

65 Dollars.

Years.	½ per ct.	1 per ct.	2 per ct.	3 per ct.	3½ per ct.	4 per ct.	4½ per ct.	5 per ct.	6 per ct.	7 per ct.	8 per ct.	10 per ct.
1	33	65	1.30	1.95	2.28	2.60	2.93	3.25	3.90	4.55	5.20	6.50
2	65	1.30	2.60	3.90	4.55	5.20	5.85	6.50	7.80	9.10	10.40	13.00
3	98	1.95	3.90	5.85	6.83	7.80	8.78	9.75	11.70	13.65	15.60	19.50
4	1.30	2.60	5.20	7.80	9.10	10.40	11.70	13.00	15.60	18.20	20.80	26.00
5	1.63	3.25	6.50	9.75	11.38	13.00	14.63	16.25	19.50	22.75	26.00	32.50

Months.												
1	03	05	11	16	19	22	24	27	33	38	43	54
2	05	11	22	33	38	43	49	54	65	76	87	1.08
3	08	16	33	49	57	65	73	81	98	1.14	1.30	1.63
4	11	22	43	65	76	87	98	1.08	1.30	1.52	1.73	2.17
5	14	27	54	81	95	1.08	1.22	1.35	1.63	1.90	2.17	2.71
6	16	33	65	98	1.14	1.30	1.46	1.63	1.95	2.28	2.60	3.25
7	19	38	76	1.14	1.33	1.52	1.71	1.90	2.28	2.65	3.03	3.79
8	22	43	87	1.30	1.52	1.73	1.95	2.17	2.60	3.03	3.47	4.33
9	24	49	98	1.46	1.71	1.95	2.19	2.44	2.93	3.41	3.90	4.88
10	27	54	1.08	1.63	1.90	2.17	2.44	2.71	3.25	3.79	4.33	5.42
11	30	60	1.19	1.79	2.09	2.38	2.68	2.98	3.58	4.17	4.77	5.96

Days.												
1	0	0	0	01	01	01	01	01	01	01	01	02
2	0	0	01	01	01	01	02	02	02	03	03	04
3	0	01	01	02	02	02	02	03	03	04	04	05
4	0	01	01	02	03	03	03	04	04	05	06	07
5	0	01	02	03	03	04	04	05	05	06	07	09
6	01	01	02	03	04	04	05	05	07	08	09	11
7	01	01	03	04	04	05	06	06	08	09	10	13
8	01	01	03	04	05	06	07	07	09	10	12	14
9	01	02	03	05	06	07	07	08	10	11	13	16
10	01	02	04	05	06	07	08	09	11	13	14	18
11	01	02	04	06	07	08	09	10	12	14	16	20
12	01	02	04	07	08	09	10	11	13	15	17	22
13	01	02	05	07	08	09	11	12	14	16	19	23
14	01	03	05	08	09	10	11	13	15	18	20	25
15	01	03	05	08	09	11	12	14	16	19	22	27
16	01	03	06	09	10	12	13	14	17	20	23	29
17	02	03	06	09	11	12	14	15	18	21	25	31
18	02	03	07	10	11	13	15	16	20	23	26	33
19	02	03	07	10	12	14	15	17	21	24	27	34
20	02	04	07	11	13	14	16	18	22	25	29	36
21	02	04	08	11	13	15	17	19	23	27	30	38
22	02	04	08	12	14	16	18	20	24	28	32	40
23	02	04	08	12	15	17	19	21	25	29	33	42
24	02	04	09	13	15	17	20	22	26	30	35	43
25	02	05	09	14	16	18	20	23	27	32	36	45
26	02	05	09	14	16	19	21	23	28	33	38	47
27	02	05	10	15	17	20	22	24	29	34	39	49
28	03	05	10	15	18	20	23	25	30	35	40	51
29	03	05	10	16	18	21	24	26	31	37	42	52
30	03	05	11	16	19	22	24	27	33	38	43	54
33	03	06	12	18	21	24	27	30	36	42	48	60
63	06	11	23	34	40	46	51	57	68	80	91	1.14
93	08	17	34	50	59	67	76	84	1.01	1.18	1.34	1.68

66 Dollars.

Years.	½ per ct.	1 per ct.	2 per ct.	3 per ct.	3½ per ct.	4 per ct.	4½ per ct.	5 per ct.	6 per ct.	7 per ct.	8 per ct.	10 per ct.
1	33	66	1.32	1.98	2.31	2.64	2.97	3.30	3.96	4.62	5.28	6.60
2	66	1.32	2.64	3.96	4.62	5.28	5.94	6.60	7.92	9.24	10.56	13.20
3	99	1.98	3.96	5.94	6.93	7.92	8.91	9.90	11.88	13.86	15.84	19.80
4	1.32	2.64	5.28	7.92	9.24	10.56	11.88	13.20	15.84	18.48	21.12	26.40
5	1.65	3.30	6.60	9.90	11.55	13.20	14.85	16.50	19.80	23.10	26.40	33.00
Months.												
1	03	06	11	17	19	22	25	28	33	39	44	55
2	06	11	22	33	39	44	50	55	66	77	88	1.10
3	08	17	33	50	58	66	74	83	99	1.16	1.32	1.65
4	11	22	44	66	77	88	99	1.10	1.32	1.54	1.76	2.20
5	14	28	55	83	96	1.10	1.24	1.38	1.65	1.93	2.20	2.75
6	17	33	66	99	1.16	1.32	1.49	1.65	1.98	2.31	2.64	3.30
7	19	39	77	1.16	1.35	1.54	1.73	1.93	2.31	2.70	3.08	3.85
8	22	44	88	1.32	1.54	1.76	1.98	2.20	2.64	3.08	3.52	4.40
9	25	50	99	1.49	1.73	1.98	2.23	2.48	2.97	3.47	3.96	4.95
10	28	55	1.10	1.65	1.93	2.20	2.48	2.75	3.30	3.85	4.40	5.50
11	30	61	1.21	1.82	2.12	2.42	2.72	3.03	3.63	4.24	4.84	6.05
Days.												
1	0	0	0	01	01	01	01	01	01	01	01	02
2	0	0	01	01	01	01	02	02	02	03	03	04
3	0	01	01	02	02	02	02	03	03	04	04	06
4	0	01	01	02	03	03	03	04	04	05	06	07
5	0	01	02	03	03	04	04	05	06	06	07	09
6	01	01	02	03	04	04	05	06	07	08	09	11
7	01	01	03	04	04	05	06	06	08	09	10	13
8	01	01	03	04	05	06	07	07	09	10	12	15
9	01	02	03	05	06	07	07	08	10	12	13	17
10	01	02	04	06	06	07	08	09	11	13	15	18
11	01	02	04	06	07	08	09	10	12	14	16	20
12	01	02	04	07	08	09	10	11	13	15	18	22
13	01	02	05	07	08	10	11	12	14	17	19	24
14	01	03	05	08	09	10	12	13	15	18	21	26
15	01	03	06	08	10	11	12	14	17	19	22	28
16	01	02	06	09	10	12	13	15	18	21	23	29
17	02	03	06	09	11	12	14	16	19	22	25	31
18	02	03	07	10	12	13	15	17	20	23	26	33
19	02	03	07	10	12	14	16	17	21	24	28	35
20	02	04	07	11	13	15	17	18	22	26	29	37
21	02	04	08	12	13	15	17	19	23	27	31	39
22	02	04	08	12	14	16	18	20	24	28	32	40
23	02	04	08	13	15	17	19	21	25	30	34	42
24	02	04	09	13	15	18	20	22	26	31	35	44
25	02	05	09	14	16	18	21	23	28	32	37	46
26	02	05	10	14	17	19	21	24	29	33	38	48
27	02	05	10	15	17	20	22	25	30	35	40	50
28	03	05	10	15	18	21	23	26	31	36	41	51
29	03	05	11	16	19	21	24	27	32	37	43	53
30	03	06	11	17	19	22	25	28	33	39	44	55
33	03	06	12	18	21	24	27	30	36	42	48	61
63	06	12	23	35	40	46	52	58	69	81	92	1.16
93	09	17	34	51	60	68	77	85	1.02	1.19	1.36	1.71

67 Dollars.

Years.	¼ per ct.	1 per ct.	2 per ct.	3 per ct.	3½ per ct.	4 per ct.	4½ per ct.	5 per ct.	6 per ct.	7 per ct.	8 per ct.	10 per ct.
1	34	67	1.34	2.01	2.35	2.68	3.02	3.35	4.02	4.69	5.36	6.70
2	67	1.34	2.68	4.02	4.69	5.36	6.03	6.70	8.04	9.38	10.72	13.40
3	1.01	2.01	4.02	6.03	7.04	8.04	9.05	10.05	12.06	14.07	16.08	20.10
4	1.34	2.68	5.36	8.04	9.38	10.72	12.06	13.40	16.08	18.76	21.44	26.80
5	1.68	3.35	6.70	10.05	11.73	13.40	15.08	16.75	20.10	23.45	26.80	33.50
Months.												
1	03	06	11	17	20	22	25	28	34	39	45	56
2	06	11	22	34	39	45	50	56	67	78	89	1.12
3	08	17	34	50	59	67	75	84	1.01	1.17	1.34	1.68
4	11	22	45	67	78	89	1.01	1.12	1.34	1.56	1.79	2.23
5	14	28	56	84	98	1.12	1.26	1.40	1.68	1.95	2.23	2.79
6	17	34	67	1.01	1.17	1.34	1.51	1.68	2.01	2.35	2.68	3.35
7	20	39	78	1.17	1.37	1.56	1.76	1.95	2.35	2.74	3.13	3.91
8	22	45	89	1.34	1.56	1.79	2.01	2.23	2.68	3.13	3.57	4.47
9	25	50	1.01	1.51	1.76	2.01	2.26	2.51	3.02	3.52	4.02	5.03
10	28	56	1.12	1.68	1.95	2.23	2.51	2.79	3.35	3.91	4.47	5.58
11	31	61	1.23	1.84	2.15	2.46	2.76	3.07	3.69	4.30	4.91	6.14
Days.												
1	0	0	0	01	01	01	01	01	01	01	01	02
2	0	0	01	01	01	01	02	02	02	03	03	04
3	0	01	01	02	02	02	03	03	03	04	04	06
4	0	01	01	02	03	03	03	04	04	05	06	07
5	0	01	02	03	03	04	04	05	06	07	07	09
6	01	01	02	03	04	04	05	06	07	08	09	11
7	01	01	03	04	05	05	06	07	08	09	10	13
8	01	01	03	04	05	06	07	07	09	10	12	15
9	01	02	03	05	06	07	08	08	10	12	13	17
10	01	02	04	06	07	07	08	09	11	13	15	19
11	01	02	04	06	07	08	09	10	12	14	16	20
12	01	02	04	07	08	09	10	11	13	16	18	22
13	01	02	05	07	08	10	11	12	15	17	19	24
14	01	03	05	08	09	10	12	13	16	18	21	26
15	01	03	06	08	10	11	13	14	17	20	22	28
16	01	03	06	09	10	12	13	15	18	21	24	30
17	02	03	06	09	11	13	14	16	19	22	25	32
18	02	03	07	10	12	13	15	17	20	23	27	34
19	02	04	07	11	12	14	16	18	21	25	28	35
20	02	04	07	11	13	15	17	19	22	26	30	37
21	02	04	08	12	14	16	18	20	23	27	31	39
22	02	04	08	12	14	16	18	20	25	29	33	41
23	02	04	09	13	15	17	19	21	26	30	34	43
24	02	04	09	13	16	18	20	22	27	31	36	45
25	02	05	09	14	16	19	21	23	28	33	37	47
26	02	05	10	15	17	19	22	24	29	34	39	48
27	03	05	10	15	18	20	23	25	30	35	40	50
28	03	05	10	16	18	21	23	26	31	36	42	52
29	03	05	11	16	19	22	24	27	32	38	43	54
30	03	06	11	17	20	22	25	28	34	39	45	56
33	03	06	12	18	21	25	28	31	37	43	49	61
63	06	12	23	35	41	47	52	59	70	82	94	1.17
93	09	17	35	52	61	69	78	87	1.04	1.21	1.38	1.73

68 Dollars.

	¼ per ct.	1 per ct.	2 per ct.	3 per ct.	3½ per ct.	4 per ct.	4½ per ct.	5 per ct.	6 per ct.	7 per ct.	8 per ct.	10 per ct.
Years.												
1	34	68	1.36	2.04	2.38	2.72	3.06	3.40	4.08	4.76	5.44	6.80
2	68	1.36	2.72	4.08	4.76	5.44	6.12	6.80	8.16	9.52	10.88	13.60
3	1.02	2.04	4.08	6.12	7.14	8.16	9.18	10.20	12.24	14.28	16.32	20.40
4	1.36	2.72	5.44	8.16	9.52	10.88	12.24	13.60	16.32	19.04	21.76	27.20
5	1.70	3.40	6.80	10.20	11.90	13.60	15.30	17.00	20.40	23.80	27.20	34.00
Months.												
1	03	06	11	17	20	23	26	28	34	40	45	57
2	06	11	23	34	40	45	51	57	68	79	91	1.13
3	09	17	34	51	60	68	77	85	1.02	1.19	1.36	1.70
4	11	23	45	68	79	91	1.02	1.13	1.36	1.59	1.81	2.27
5	14	28	57	85	99	1.13	1.28	1.42	1.70	1.98	2.27	2.83
6	17	34	68	1.02	1.19	1.36	1.53	1.70	2.04	2.38	2.72	3.40
7	20	40	79	1.19	1.39	1.59	1.79	1.98	2.38	2.78	3.17	3.97
8	23	45	91	1.36	1.59	1.81	2.04	2.27	2.72	3.17	3.63	4.53
9	26	51	1.02	1.53	1.79	2.04	2.30	2.55	3.06	3.57	4.08	5.10
10	28	57	1.13	1.70	1.98	2.27	2.55	2.83	3.40	3.97	4.53	5.67
11	31	62	1.25	1.87	2.18	2.49	2.81	3.12	3.74	4.36	4.99	6.23
Days.												
1	0	0	0	01	01	01	01	01	01	01	02	02
2	0	0	01	01	01	02	02	02	02	03	03	04
3	0	01	01	02	02	02	03	03	03	04	05	06
4	0	01	02	02	03	03	03	04	05	05	06	08
5	0	01	02	03	03	04	04	05	06	07	08	09
6	01	01	02	03	04	05	05	06	07	08	09	11
7	01	01	03	04	05	05	06	07	08	09	11	13
8	01	02	03	05	05	06	07	08	09	11	12	15
9	01	02	03	05	06	07	08	09	10	12	14	17
10	01	02	04	06	07	08	09	09	11	13	15	19
11	01	02	04	06	07	08	09	10	12	15	17	21
12	01	02	05	07	08	09	10	11	14	16	18	23
13	01	02	05	07	09	10	11	12	15	17	20	25
14	01	03	05	08	09	11	12	13	16	19	21	26
15	01	03	06	09	10	11	13	14	17	20	23	28
16	02	03	06	09	11	12	14	15	18	21	24	30
17	02	03	06	10	11	13	14	16	19	22	26	32
18	02	03	07	10	12	14	15	17	20	24	27	34
19	02	04	07	11	13	14	16	18	22	25	29	36
20	02	04	08	11	13	15	17	19	23	26	30	38
21	02	04	08	12	14	16	18	20	24	28	32	40
22	02	04	08	12	15	17	19	21	25	29	33	42
23	02	04	09	13	15	17	20	22	26	30	35	43
24	02	05	09	14	16	18	20	23	27	32	36	45
25	02	05	09	14	17	19	21	24	28	33	38	47
26	02	05	10	15	17	20	22	25	29	34	39	49
27	03	05	10	15	18	20	23	26	31	36	41	51
28	03	05	11	16	19	21	24	26	32	37	42	53
29	03	05	11	16	19	22	25	27	33	38	44	55
30	03	06	11	17	20	23	26	28	34	40	45	57
33	03	06	12	19	22	25	28	31	37	44	50	62
63	06	12	24	36	42	48	54	60	71	83	95	1.19
93	09	18	35	53	61	70	79	88	1.05	1.23	1.41	1.76

69 Dollars.

Years.	½ per ct.	1 per ct.	2 per ct.	3 per ct.	3½ per ct.	4 per ct.	4½ per ct.	5 per ct.	6 per ct.	7 per ct.	8 per ct.	10 per ct.
1	35	69	1.38	2.07	2.42	2.76	3.11	3.45	4.14	4.83	5.52	6.90
2	69	1.38	2.76	4.14	4.83	5.52	6.21	6.90	8.28	9.66	11.04	13.80
3	1.04	2.07	4.14	6.21	7.25	8.28	9.32	10.35	12.42	14.49	16.56	20.70
4	1.38	2.76	5.52	8.28	9.66	11.04	12.42	13.80	16.56	19.32	22.08	27.60
5	1.73	3.45	6.90	10.35	12.08	13.80	15.53	17.25	20.70	24.15	27.60	34.50
Months.												
1	03	06	12	17	20	23	26	29	35	40	46	58
2	06	12	23	35	40	46	52	58	69	81	92	1.15
3	09	17	35	52	60	69	78	86	1.04	1.21	1.38	1.73
4	12	23	46	69	81	92	1.04	1.15	1.38	1.61	1.84	2.30
5	14	29	58	86	1.01	1.15	1.29	1.44	1.73	2.01	2.30	2.88
6	17	35	69	1.04	1.21	1.38	1.55	1.73	2.07	2.42	2.76	3.45
7	20	40	81	1.21	1.41	1.61	1.81	2.01	2.42	2.82	3.22	4.03
8	23	46	92	1.38	1.61	1.84	2.07	2.30	2.76	3.22	3.68	4.60
9	26	52	1.04	1.55	1.81	2.07	2.33	2.59	3.11	3.62	4.14	5.18
10	29	58	1.15	1.73	2.01	2.30	2.59	2.88	3.45	4.03	4.60	5.75
11	32	63	1.27	1.90	2.21	2.53	2.85	3.16	3.80	4.43	5.06	6.33
Days.												
1	0	0	0	01	01	01	01	01	01	01	02	02
2	0	0	01	01	01	02	02	02	02	03	03	04
3	0	01	01	02	02	02	03	03	03	04	05	06
4	0	01	02	02	03	03	03	04	05	05	06	08
5	0	01	02	03	03	04	04	05	06	07	08	10
6	01	01	02	03	04	05	05	06	07	08	09	12
7	01	01	03	04	05	05	06	07	08	09	11	13
8	01	02	03	05	05	06	07	08	09	11	12	15
9	01	02	03	05	06	07	08	09	10	12	14	17
10	01	02	04	06	07	08	09	10	12	13	15	19
11	01	02	04	06	07	08	09	11	13	15	17	21
12	01	02	05	07	08	09	10	12	14	16	18	23
13	01	02	05	07	09	10	11	12	15	17	20	25
14	01	03	05	08	09	11	12	13	16	19	21	27
15	01	03	06	09	10	12	13	14	17	20	23	29
16	02	03	06	09	11	12	14	15	18	21	25	31
17	02	03	07	10	11	13	15	16	20	23	26	33
18	02	03	07	10	12	14	16	17	21	24	28	35
19	02	04	07	11	13	15	16	18	22	25	29	36
20	02	04	08	12	13	15	17	19	23	27	31	38
21	02	04	08	12	14	16	18	20	24	28	32	40
22	02	04	08	13	15	17	19	21	25	30	34	42
23	02	04	09	13	15	18	20	22	26	31	35	44
24	02	05	09	14	16	18	21	23	28	32	37	46
25	02	05	10	14	17	19	22	24	29	34	38	48
26	02	05	10	15	17	20	22	25	30	35	40	50
27	03	05	10	16	18	21	23	26	31	36	41	52
28	03	05	11	16	19	21	24	27	32	38	43	54
29	03	06	11	17	19	22	25	28	33	39	44	56
30	03	06	12	17	20	23	26	29	35	40	46	58
33	03	06	13	19	22	25	28	32	38	44	51	63
63	06	12	24	36	42	48	54	60	72	84	97	1.21
93	09	18	36	53	62	71	80	89	1.07	1.25	1.43	1.78

70 Dollars.

	½ per ct.	1 per ct.	2 per ct.	3 per ct.	3½ per ct.	4 per ct.	4½ per ct.	5 per ct.	6 per ct.	7 per ct.	8 per ct.	10 per ct.
Years.												
1	35	70	1.40	2.10	2.45	2.80	3.15	3.50	4.20	4.90	5.60	7.00
2	70	1.40	2.80	4.20	4.90	5.60	6.30	7.00	8.40	9.80	11.20	14.00
3	1.05	2.10	4.20	6.30	7.35	8.40	9.45	10.50	12.60	14.70	16.80	21.00
4	1.40	2.80	5.60	8.40	9.80	11.20	12.60	14.00	16.80	19.60	22.40	28.00
5	1.75	3.50	7.00	10.50	12.25	14.00	15.75	17.50	21.00	24.50	28.00	35.00
Months.												
1	03	06	12	18	20	23	26	29	35	41	47	58
2	06	12	23	35	41	47	53	58	70	82	93	1.17
3	09	18	35	53	61	70	79	88	1.05	1.23	1.40	1.75
4	12	23	47	70	82	93	1.05	1.17	1.40	1.63	1.87	2.33
5	15	29	58	88	1.02	1.17	1.31	1.46	1.75	2.04	2.33	2.92
6	18	35	70	1.05	1.23	1.40	1.58	1.75	2.10	2.45	2.80	3.50
7	20	41	82	1.23	1.43	1.63	1.84	2.04	2.45	2.86	3.27	4.08
8	23	47	93	1.40	1.63	1.87	2.10	2.33	2.80	3.27	3.73	4.67
9	26	53	1.05	1.58	1.84	2.10	2.36	2.63	3.15	3.68	4.20	5.25
10	29	58	1.17	1.75	2.04	2.33	2.63	2.92	3.50	4.08	4.67	5.83
11	32	64	1.28	1.93	2.25	2.57	2.89	3.21	3.85	4.49	5.13	6.42
Days.												
1	0	0	0	01	01	01	01	01	01	01	02	02
2	0	0	01	01	01	02	02	02	02	03	03	04
3	0	01	01	02	02	02	03	03	04	04	05	06
4	0	01	02	02	03	03	04	04	05	05	06	08
5	0	01	02	03	03	04	04	05	06	07	08	10
6	01	01	02	04	04	05	05	06	07	08	09	12
7	01	01	03	04	05	05	06	07	08	10	11	14
8	01	02	03	05	05	06	07	08	09	11	12	16
9	01	02	04	05	06	07	08	09	11	12	14	18
10	01	02	04	06	07	08	09	10	12	14	16	19
11	01	02	04	06	07	09	10	11	13	15	17	21
12	01	02	05	07	08	09	11	12	14	16	19	23
13	01	03	05	08	09	10	11	13	15	18	20	25
14	01	03	05	08	10	11	12	14	16	19	22	27
15	01	03	06	09	10	12	13	15	18	20	23	29
16	02	03	06	09	11	12	14	16	19	22	25	31
17	02	03	07	10	12	13	15	17	20	23	26	33
18	02	04	07	11	12	14	16	18	21	25	28	35
19	02	04	07	11	13	15	17	18	22	26	30	37
20	02	04	08	12	14	16	18	19	23	27	31	39
21	02	04	08	12	14	16	18	20	25	29	33	41
22	02	04	09	13	15	17	19	21	26	30	34	43
23	02	04	09	13	16	18	20	22	27	31	36	45
24	02	05	09	14	16	19	21	23	28	33	37	47
25	02	05	10	15	17	19	22	24	29	34	39	49
26	03	05	10	15	18	20	23	25	30	35	40	51
27	03	05	11	16	18	21	24	26	32	37	42	53
28	03	05	11	16	19	22	25	27	33	38	44	54
29	03	06	11	17	20	23	25	28	34	39	45	56
30	03	06	12	18	20	23	26	29	35	41	47	58
33	03	06	13	19	22	26	29	32	38	45	51	64
63	06	12	25	37	43	49	55	61	74	86	98	1.23
93	09	18	36	54	63	72	81	90	1.09	1.27	1.45	1.81

71 Dollars.

Years.	½ per ct.	1 per ct.	2 per ct.	3 per ct.	3½ per ct.	4 per ct.	4½ per ct.	5 per ct.	6 per ct.	7 per ct.	8 per ct.	10 per ct.
1	36	71	1.42	2.13	2.49	2.84	3.20	3.55	4.26	4.97	5.68	7.10
2	71	1.42	2.84	4.26	4.97	5.68	6.39	7.10	8.52	9.94	11.36	14.20
3	1.07	2.13	4.26	6.39	7.46	8.52	9.59	10.65	12.78	14.91	17.04	21.30
4	1.42	2.84	5.68	8.52	9.94	11.36	12.78	14.20	17.04	19.88	22.72	28.40
5	1.78	3.55	7.10	10.65	12.43	14.20	15.98	17.75	21.30	24.85	28.40	35.50

Months.												
1	03	06	12	18	21	24	27	30	36	41	47	59
2	06	12	24	36	41	47	53	59	71	83	95	1.18
3	09	18	36	53	62	71	80	89	1.07	1.24	1.42	1.78
4	12	24	47	71	83	95	1.07	1.18	1.42	1.66	1.89	2.37
5	15	30	59	89	1.04	1.18	1.33	1.48	1.78	2.07	2.37	2.96
6	18	36	71	1.07	1.24	1.42	1.60	1.78	2.13	2.49	2.84	3.55
7	21	41	83	1.24	1.45	1.66	1.86	2.07	2.49	2.90	3.31	4.14
8	24	47	95	1.42	1.66	1.89	2.13	2.37	2.84	3.31	3.79	4.73
9	27	53	1.07	1.60	1.86	2.13	2.40	2.66	3.20	3.73	4.26	5.33
10	30	59	1.18	1.78	2.07	2.37	2.66	2.96	3.55	4.14	4.73	5.92
11	33	65	1.30	1.95	2.28	2.60	2.93	3.25	3.91	4.56	5.21	6.51

Days.												
1	0	0	0	01	01	01	01	01	01	01	02	02
2	0	0	01	01	01	02	02	02	02	03	03	04
3	0	01	01	02	02	02	03	03	04	04	05	06
4	0	01	02	02	03	03	04	04	05	05	06	08
5	0	01	02	03	03	04	04	05	06	07	08	10
6	01	01	02	04	04	05	05	06	07	08	09	12
7	01	01	03	04	05	06	06	07	08	10	11	14
8	01	02	03	05	06	06	07	08	09	11	13	16
9	01	02	04	05	06	07	08	09	11	12	14	18
10	01	02	04	06	07	08	09	10	12	14	16	20
11	01	02	04	07	08	09	10	11	13	15	17	22
12	01	02	05	07	08	09	11	12	14	17	19	24
13	01	03	05	08	09	10	12	13	15	18	21	26
14	01	03	06	08	10	11	12	14	17	19	22	28
15	01	03	06	09	10	12	13	15	18	21	24	30
16	02	03	06	09	11	13	14	16	19	22	25	32
17	02	03	07	10	12	13	15	17	20	23	27	34
18	02	04	07	11	12	14	16	18	21	25	28	36
19	02	04	07	11	13	15	17	19	22	26	30	37
20	02	04	08	12	14	16	18	20	24	28	32	39
21	02	04	08	12	14	17	19	21	25	29	33	41
22	02	04	09	13	15	17	20	22	26	30	35	43
23	02	05	09	14	16	18	20	23	27	32	36	45
24	02	05	09	14	17	19	21	24	28	33	38	47
25	02	05	10	15	17	20	22	25	30	35	39	49
26	03	05	10	15	18	21	23	26	31	36	41	51
27	03	05	11	16	19	21	24	27	32	37	43	53
28	03	06	11	17	19	22	25	28	33	39	44	55
29	03	06	11	17	20	23	26	29	34	40	46	57
30	03	06	12	18	21	24	27	30	36	41	47	59
33	03	07	13	20	23	26	29	33	39	46	52	65
63	06	12	25	37	43	50	56	62	75	87	99	1.24
93	09	18	37	55	64	73	83	92	1.10	1.28	1.47	1.83

72 Dollars.

	½ per ct.	1 per ct.	2 per ct.	3 per ct.	3½ per ct.	4 per ct.	4½ per ct.	5 per ct.	6 per ct.	7 per ct.	8 per ct.	10 per ct.
Years.												
1	36	72	1.44	2.16	2.52	2.88	3.24	3.60	4.32	5.04	5.76	7.20
2	72	1.44	2.88	4.32	5.04	5.76	6.48	7.20	8.64	10.08	11.52	14.40
3	1.08	2.16	4.32	6.48	7.56	8.64	9.72	10.80	12.96	15.12	17.28	21.60
4	1.44	2.88	5.76	8.64	10.08	11.52	12.96	14.40	17.28	20.16	23.04	28.80
5	1.80	3.60	7.20	10.80	12.60	14.40	16.20	18.00	21.60	25.20	28.80	36.00
Months.												
1	03	06	12	18	21	24	27	30	36	42	48	60
2	06	12	24	36	42	48	54	60	72	84	96	1.20
3	09	18	36	54	63	72	81	90	1.08	1.26	1.44	1.80
4	12	24	48	72	84	96	1.08	1.20	1.44	1.68	1.92	2.40
5	15	30	60	90	1.05	1.20	1.35	1.50	1.80	2.10	2.40	3.00
6	18	36	72	1.08	1.26	1.44	1.62	1.80	2.16	2.52	2.88	3.60
7	21	42	84	1.26	1.47	1.68	1.89	2.10	2.52	2.94	3.36	4.20
8	24	48	96	1.44	1.68	1.92	2.16	2.40	2.88	3.36	3.84	4.80
9	27	54	1.08	1.62	1.89	2.16	2.43	2.70	3.24	3.78	4.32	5.40
10	30	60	1.20	1.80	2.10	2.40	2.70	3.00	3.60	4.20	4.80	6.00
11	33	66	1.32	1.98	2.31	2.64	2.97	3.30	3.96	4.62	5.28	6.60
Days.												
1	0	0	0	01	01	01	01	01	01	01	02	02
2	0	0	01	01	01	02	02	02	02	03	03	04
3	0	01	01	02	02	02	03	03	04	04	05	06
4	0	01	02	02	03	03	04	04	05	06	06	08
5	01	01	02	03	04	04	05	05	06	07	08	10
6	01	01	02	04	04	05	05	06	07	08	10	12
7	01	01	03	04	05	06	06	07	08	10	11	14
8	01	02	03	05	06	06	07	08	10	11	13	16
9	01	02	04	05	06	07	08	09	11	13	14	18
10	01	02	04	06	07	08	09	10	12	14	16	20
11	01	02	04	07	08	09	10	11	13	15	18	22
12	01	02	05	07	08	10	11	12	14	17	19	24
13	01	03	05	08	09	10	12	13	16	18	21	26
14	01	03	06	08	10	11	13	14	17	20	22	28
15	02	03	06	09	11	12	14	15	18	21	24	30
16	02	03	06	10	11	13	14	16	19	22	26	32
17	02	03	07	10	12	14	15	17	20	24	27	34
18	02	04	07	11	13	14	16	18	22	25	29	36
19	02	04	08	11	13	15	17	19	23	27	30	38
20	02	04	08	12	14	16	18	20	24	28	32	40
21	02	04	08	13	15	17	19	21	25	29	34	42
22	02	04	09	13	15	18	20	22	26	31	35	44
23	02	05	09	14	16	18	21	23	28	32	37	46
24	02	05	10	14	17	19	22	24	29	34	38	48
25	03	05	10	15	18	20	23	25	30	35	40	50
26	03	05	10	16	18	21	23	26	31	36	42	52
27	03	05	11	16	19	22	24	27	32	38	43	54
28	03	06	11	17	20	22	25	28	34	39	45	56
29	03	06	12	17	20	23	26	29	35	41	46	58
30	03	06	12	18	21	24	27	30	36	42	48	60
33	03	07	13	20	23	26	30	33	40	46	53	66
63	06	13	25	38	44	50	57	63	76	88	1.01	1.26
93	09	19	37	56	65	74	84	93	1.12	1.30	1.49	1.86

73 Dollars.

Years.	½ per ct.	1 per ct.	2 per ct.	3 per ct.	3½ per ct.	4 per ct.	4½ per ct.	5 per ct.	6 per ct.	7 per ct.	8 per ct.	10 per ct.
1	37	73	1.46	2.19	2.56	2.92	3.29	3.65	4.38	5.11	5.84	7.30
2	73	1.46	2.92	4.38	5.11	5.84	6.57	7.30	8.76	10.22	11.68	14.60
3	1.10	2.19	4.38	6.57	7.67	8.76	9.86	10.95	13.14	15.33	17.52	21.90
4	1.46	2.92	5.84	8.76	10.22	11.68	13.14	14.60	17.52	20.44	23.36	29.20
5	1.83	3.65	7.30	10.95	12.78	14.60	16.43	18.25	21.90	25.55	29.20	36.50
Months.												
1	03	06	12	18	21	24	27	30	37	43	.49	61
2	06	12	24	37	43	49	55	61	73	85	97	1.22
3	09	18	37	55	64	73	82	91	1.10	1.28	1.46	1.83
4	12	24	49	73	85	97	1.10	1.22	1.46	1.70	1.95	2.43
5	15	30	61	91	1.06	1.22	1.37	1.52	1.83	2.13	2.43	3.04
6	18	37	73	1.10	1.28	1.46	1.64	1.83	2.19	2.56	2.92	3.65
7	21	43	85	1.28	1.49	1.70	1.92	2.13	2.56	2.98	3.41	4.26
8	24	49	97	1.46	1.70	1.95	2.19	2.43	2.92	3.41	3.89	4.87
9	27	55	1.10	1.64	1.92	2.19	2.46	2.74	3.29	3.83	4.38	5.48
10	30	61	1.22	1.83	2.13	2.43	2.74	3.04	3.65	4.26	4.87	6.08
11	33	67	1.34	2.01	2.34	2.68	3.01	3.35	4.02	4.68	5.35	6.69
Days.												
1	0	0	0	01	01	01	01	01	01	01	02	02
2	0	0	01	01	01	02	02	02	02	03	03	04
3	0	01	01	02	02	02	03	03	04	04	05	06
4	0	01	02	02	03	03	04	04	05	06	06	08
5	01	01	02	03	04	04	05	05	06	07	08	10
6	01	01	02	04	04	05	05	06	07	09	10	12
7	01	01	03	04	05	06	06	07	09	10	11	14
8	01	02	03	05	06	06	07	08	10	11	13	16
9	01	02	04	05	06	07	08	09	11	13	15	18
10	01	02	04	06	07	08	09	10	12	14	16	20
11	01	02	04	07	08	09	10	11	13	16	18	22
12	01	02	05	07	09	10	11	12	15	17	19	24
13	01	03	05	08	09	11	12	13	16	18	21	26
14	01	03	06	09	10	11	13	14	17	20	23	28
15	02	03	06	09	11	12	14	15	18	21	24	30
16	02	03	06	10	11	13	15	16	19	23	26	32
17	02	03	07	10	12	14	16	17	21	24	28	34
18	02	04	07	11	13	15	16	18	22	26	29	37
19	02	04	08	12	13	15	17	19	23	27	31	39
20	02	04	08	12	14	16	18	20	24	28	32	41
21	02	04	09	13	15	17	19	21	26	30	34	43
22	02	04	09	13	16	18	20	22	27	31	36	45
23	02	05	09	14	16	19	21	23	28	33	37	47
24	02	05	10	15	17	19	22	24	29	34	39	49
25	03	05	10	15	18	20	23	25	30	35	41	51
26	03	05	11	16	18	21	24	26	32	37	42	53
27	03	05	11	16	19	22	25	27	33	38	44	55
28	03	06	11	17	20	23	26	28	34	40	45	57
29	03	06	12	18	21	24	26	29	35	41	47	59
30	03	06	12	18	21	24	27	30	37	43	49	61
33	03	07	13	20	23	27	30	33	40	47	54	67
63	06	13	26	38	45	51	57	64	77	89	1.02	1.28
93	09	19	38	56	66	75	84	94	1.13	1.32	1.50	1.88

74 Dollars.

	½ per ct.	1 per ct.	2 per ct.	3 per ct.	3½ per ct.	4 per ct.	4½ per ct.	5 per ct.	6 per ct.	7 per ct.	8 per ct.	10 per ct.
Years.												
1	37	74	1.48	2.22	2.59	2.96	3.33	3.70	4.44	5.18	5.92	7.40
2	74	1.48	2.96	4.44	5.18	5.92	6.66	7.40	8.88	10.36	11.84	14.80
3	1.11	2.22	4.44	6.66	7.77	8.88	9.99	11.10	13.32	15.54	17.76	22.20
4	1.48	2.96	5.92	8.88	10.36	11.84	13.32	14.80	17.76	20.72	23.68	29.60
5	1.85	3.70	7.40	11.10	12.95	14.80	16.65	18.50	22.20	25.90	29.60	37.00
Months.												
1	03	06	12	19	22	25	28	31	37	43	49	62
2	06	12	25	37	43	49	56	62	74	86	99	1.23
3	09	19	37	56	65	74	83	93	1.11	1.30	1.48	1.85
4	12	24	49	74	86	99	1.11	1.23	1.48	1.73	1.97	2.47
5	15	31	62	93	1.08	1.23	1.39	1.54	1.85	2.16	2.47	3.08
6	19	37	74	1.11	1.30	1.48	1.67	1.85	2.22	2.59	2.96	3.70
7	22	43	86	1.30	1.51	1.73	1.94	2.16	2.59	3.02	3.45	4.32
8	25	49	99	1.48	1.73	1.97	2.22	2.47	2.96	3.45	3.95	4.93
9	28	56	1.11	1.67	1.94	2.22	2.50	2.78	3.33	3.89	4.44	5.55
10	31	62	1.23	1.85	2.16	2.47	2.78	3.08	3.70	4.32	4.93	6.17
11	34	68	1.36	2.04	2.37	2.71	3.05	3.39	4.07	4.75	5.43	6.78
Days.												
1	0	0	0	01	01	01	01	01	01	01	02	02
2	0	0	01	01	01	02	02	02	02	03	03	04
3	0	01	01	02	02	02	03	03	04	04	05	06
4	0	01	02	02	03	03	04	04	05	06	07	08
5	01	01	02	03	04	04	05	05	06	07	08	10
6	01	01	02	04	04	05	06	06	07	09	10	12
7	01	01	03	04	05	06	06	07	09	10	12	14
8	01	02	03	05	06	07	07	08	10	12	13	16
9	01	02	04	06	06	07	08	09	11	13	15	19
10	01	02	04	06	07	08	09	10	12	14	16	21
11	01	02	05	07	08	09	10	11	14	16	18	23
12	01	02	05	07	09	10	11	12	15	17	20	25
13	01	03	05	08	09	11	12	13	16	19	21	27
14	01	03	06	09	10	12	13	14	17	20	23	29
15	02	03	06	09	11	12	14	15	19	22	25	31
16	02	03	07	10	12	13	15	16	20	23	26	33
17	02	03	07	10	12	14	16	17	21	24	28	35
18	02	04	07	11	13	15	17	19	22	26	30	37
19	02	04	08	12	14	16	18	20	23	27	31	39
20	02	04	08	12	14	16	19	21	25	29	33	41
21	02	04	09	13	15	17	19	22	26	30	35	43
22	02	05	09	14	16	18	20	23	27	32	36	45
23	02	05	09	14	17	19	21	24	28	33	38	47
24	02	05	10	15	17	20	22	25	30	35	39	49
25	03	05	10	15	18	21	23	26	31	36	41	51
26	03	05	11	16	19	21	24	27	32	37	43	53
27	03	06	11	17	19	22	25	28	33	39	44	56
28	03	06	12	17	20	23	26	29	35	40	46	58
29	03	06	12	18	21	24	27	30	36	42	48	60
30	03	06	12	19	22	25	28	31	37	43	49	62
33	03	07	14	20	24	27	31	34	41	47	54	68
63	06	13	26	39	45	52	58	65	78	91	1.04	1.30
93	10	19	38	57	67	76	86	96	1.15	1.34	1.53	1.91

75 Dollars.

Years.	½ per ct.	1 per ct.	2 per ct.	3 per ct.	3½ per ct.	4 per ct.	4½ per ct.	5 per ct.	6 per ct.	7 per ct.	8 per ct.	10 per ct.
1	38	75	1.50	2.25	2.63	3.00	8.38	3.75	4.50	5.25	6.00	7.50
2	75	1.50	3.00	4.50	5.25	6.00	6.75	7.50	9.00	10.50	12.00	15.00
3	1.13	2.25	4.50	6.75	7.87	9.00	10.13	11.25	13.50	15.75	18.00	22.50
4	1.50	3.00	6.00	9.00	10.50	12.00	13.50	15.00	18.00	21.00	24.00	30.00
5	1.88	3.75	7.50	11.25	13.13	15.00	16.88	18.75	22.50	26.25	30.00	37.50
Months.												
1	03	06	13	19	22	25	28	31	38	44	50	63
2	06	13	25	38	44	50	56	63	75	88	1.00	1.25
3	09	19	38	56	66	75	84	94	1.13	1.31	1.50	1.88
4	13	25	50	75	88	1.00	1.13	1.25	1.50	1.75	2.00	2.50
5	16	31	63	94	1.09	1.25	1.41	1.56	1.88	2.19	2.50	3.13
6	19	38	75	1.13	1.31	1.50	1.69	1.88	2.25	2.63	3.00	3.75
7	22	44	88	1.31	1.53	1.75	1.97	2.19	2.63	3.06	3.50	4.38
8	25	50	1.00	1.50	1.75	2.00	2.25	2.50	3.00	3.50	4.00	5.00
9	28	56	1.13	1.69	1.97	2.25	2.53	2.81	3.38	3.94	4.50	5.63
10	31	63	1.25	1.88	2.19	2.50	2.81	3.13	3.75	4.38	5.00	6.25
11	34	69	1.38	2.06	2.41	2.75	3.09	3.44	4.13	4.81	5.50	6.88
Days.												
1	0	0	0	01	01	01	01	01	01	01	02	02
2	0	0	01	01	01	02	02	02	03	03	03	04
3	0	01	01	02	02	03	03	03	04	04	05	06
4	0	01	02	03	03	03	04	04	05	06	07	08
5	01	01	02	03	04	04	05	05	06	07	08	10
6	01	01	03	04	04	05	06	06	08	09	10	13
7	01	01	03	04	05	06	07	07	09	10	12	15
8	01	02	03	05	06	07	08	08	10	12	13	17
9	01	02	04	06	07	08	08	09	11	13	15	19
10	01	02	04	06	07	08	09	10	13	15	17	21
11	01	02	05	07	08	09	10	11	14	16	18	23
12	01	03	05	08	09	10	11	13	15	18	20	25
13	01	03	05	08	09	11	12	14	16	19	22	27
14	01	03	06	09	10	12	13	15	18	20	23	29
15	02	03	06	09	11	13	14	16	19	22	25	31
16	02	03	07	10	12	13	15	17	20	23	27	33
17	02	04	07	11	12	14	16	18	21	25	28	35
18	02	04	08	11	13	15	17	19	23	26	30	38
19	02	04	08	12	14	16	18	20	24	28	32	40
20	02	04	08	13	15	17	19	21	25	29	33	42
21	02	04	09	13	15	18	20	22	26	31	35	44
22	02	05	09	14	16	18	21	23	28	32	37	46
23	02	05	10	14	17	19	22	24	29	34	38	48
24	03	05	10	15	18	20	23	25	30	35	40	50
25	03	05	10	16	18	21	23	26	31	36	42	52
26	03	05	11	16	19	22	24	27	33	38	43	54
27	03	06	11	17	20	23	25	28	34	39	45	56
28	03	06	12	18	20	23	26	29	35	41	47	58
29	03	06	12	18	21	24	27	30	36	42	48	60
30	03	06	13	19	22	25	28	31	38	44	50	63
33	03	07	14	21	24	28	31	34	41	48	55	69
63	07	13	26	39	46	53	59	66	79	92	1.05	1.31
93	10	19	39	58	68	78	87	97	1.16	1.36	1.55	1.94

76 Dollars.

	½ per ct.	1 per ct.	2 per ct.	3 per ct.	3½ per ct.	4 per ct.	4½ per ct.	5 per ct.	6 per ct.	7 per ct.	8 per ct.	10 per ct.
Years.												
1	38	76	1.52	2.28	2.66	3.04	3.42	3.80	4.56	5.32	6.08	7.60
2	76	1.52	3.04	4.56	5.32	6.08	6.84	7.60	9.12	10.64	12.16	15.20
3	1.14	2.28	4.56	6.84	7.98	9.12	10.26	11.40	13.68	15.96	18.24	22.80
4	1.52	3.04	6.08	9.12	10.64	12.16	13.68	15.20	18.24	21.28	24.32	30.40
5	1.90	3.80	7.60	11.40	13.30	15.20	17.10	19.00	22.80	26.60	30.40	38.00
Months.												
1	03	06	13	19	22	25	29	32	38	44	51	63
2	06	13	25	38	44	51	57	63	76	89	1.01	1.27
3	10	19	38	57	67	76	86	95	1.14	1.33	1.52	1.90
4	13	25	51	76	89	1.01	1.14	1.27	1.52	1.77	2.03	2.53
5	16	33	63	95	1.11	1.27	1.43	1.58	1.90	2.22	2.53	3.17
6	19	38	76	1.14	1.33	1.52	1.71	1.90	2.28	2.66	3.04	3.80
7	22	44	89	1.33	1.55	1.77	2.00	2.22	2.66	3.10	3.55	4.43
8	25	51	1.01	1.52	1.77	2.03	2.28	2.53	3.04	3.55	4.05	5.07
9	29	57	1.14	1.71	2.00	2.28	2.57	2.85	3.42	3.99	4.56	5.70
10	32	63	1.27	1.90	2.22	2.53	2.85	3.17	3.80	4.43	5.07	6.33
11	35	70	1.31	2.09	2.44	2.79	3.14	3.48	4.18	4.88	5.57	6.97
Days.												
1	0	0	0	01	01	01	01	01	01	01	02	02
2	0	0	01	01	01	02	02	02	03	03	03	04
3	0	01	01	02	02	03	03	03	04	04	05	06
4	0	01	02	03	03	03	04	04	05	06	07	08
5	01	01	02	03	04	04	05	05	06	07	08	11
6	01	01	03	04	04	05	06	06	08	09	10	13
7	01	01	03	04	05	06	07	07	09	10	12	15
8	01	02	03	05	06	07	08	08	10	12	14	17
9	01	02	04	06	07	08	09	10	11	13	15	19
10	01	02	04	06	07	08	10	11	13	15	17	21
11	01	02	05	07	08	09	10	12	14	16	19	23
12	01	03	05	08	09	10	11	13	15	18	20	25
13	01	03	05	08	10	11	12	14	16	19	22	27
14	01	03	06	09	10	12	13	15	18	21	24	30
15	02	03	06	10	11	13	14	16	19	22	25	32
16	02	03	07	10	12	14	15	17	20	24	27	34
17	02	04	07	11	13	14	16	18	22	25	29	36
18	02	04	08	11	13	15	17	19	23	27	30	38
19	02	04	08	12	14	16	18	20	24	28	32	40
20	02	04	08	13	15	17	19	21	25	30	34	42
21	02	04	09	13	16	18	20	22	27	31	35	44
22	02	05	09	14	16	19	21	23	28	33	37	46
23	02	05	10	15	17	19	22	24	29	34	39	49
24	03	05	10	15	18	20	23	25	30	35	41	51
25	03	05	11	16	18	21	24	26	32	37	42	53
26	03	05	11	16	19	22	25	27	33	38	44	55
27	03	06	11	17	20	23	26	29	34	40	46	57
28	03	06	12	18	21	24	27	30	35	41	47	59
29	03	06	12	18	21	24	28	31	37	43	49	61
30	03	06	13	19	22	25	29	32	38	44	51	63
33	03	07	14	21	24	28	31	35	42	49	56	70
63	07	13	27	40	47	53	60	67	80	93	1.06	1.33
93	10	20	39	59	69	79	88	98	1.18	1.37	1.57	1.96

77 Dollars.

Years.	½ per ct.	1 per ct.	2 per ct.	3 per ct.	3½ per ct.	4 per ct.	4½ per ct.	5 per ct.	6 per ct.	7 per ct.	8 per ct.	10 per ct.
1	39	77	1.54	2.31	2.70	3.08	3.47	3.85	4.62	5.39	6.16	7.70
2	77	1.54	3.08	4.62	5.39	6.16	6.93	7.70	9.24	10.78	12.32	15.40
3	1.16	2.31	4.62	6.93	8.09	9.24	10.40	11.55	13.86	16.17	18.48	23.10
4	1.54	3.08	6.16	9.24	10.78	12.32	13.86	15.40	18.48	21.56	24.64	30.80
5	1.93	3.85	7.70	11.55	13.48	15.40	17.33	19.25	23.10	26.95	30.80	38.50
Months.												
1	03	06	13	19	22	26	29	32	39	45	51	64
2	06	13	26	39	45	51	58	64	77	90	1.03	1.28
3	10	19	39	58	67	77	87	96	1.16	1.35	1.54	1.93
4	13	26	51	77	90	1.03	1.16	1.28	1.54	1.80	2.05	2.57
5	16	32	64	96	1.12	1.28	1.44	1.60	1.93	2.25	2.57	3.21
6	19	39	77	1.16	1.35	1.54	1.73	1.93	2.31	2.70	3.08	3.85
7	22	45	90	1.35	1.57	1.80	2.02	2.25	2.70	3.14	3.59	4.49
8	26	51	1.03	1.54	1.80	2.05	2.31	2.57	3.08	3.59	4.11	5.13
9	29	58	1.16	1.73	2.02	2.31	2.60	2.89	3.47	4.04	4.62	5.78
10	32	64	1.28	1.93	2.25	2.57	2.89	3.21	3.85	4.49	5.13	6.42
11	35	71	1.41	2.12	2.47	2.82	3.18	3.53	4.24	4.94	5.65	7.06
Days.												
1	0	0	0	01	01	01	01	01	01	01	02	02
2	0	0	01	01	01	02	02	02	03	03	03	04
3	0	01	01	02	02	03	03	03	04	04	05	06
4	0	01	02	03	03	03	04	04	05	06	07	09
5	01	01	02	03	04	04	05	05	06	07	09	11
6	01	01	03	04	04	05	06	06	08	09	10	13
7	01	01	03	04	05	06	07	07	09	10	12	15
8	01	02	03	05	06	07	08	09	10	12	14	17
9	01	02	04	06	07	08	09	10	12	13	15	19
10	01	02	04	06	07	09	10	11	13	15	17	21
11	01	02	05	07	08	09	11	12	14	16	19	24
12	01	03	05	08	09	10	12	13	15	18	21	26
13	01	03	06	08	10	11	13	14	17	19	22	28
14	01	03	06	09	10	12	13	15	18	21	24	30
15	02	03	06	10	11	13	14	16	19	22	26	32
16	02	03	07	10	12	14	15	17	21	24	27	34
17	02	04	07	11	13	15	16	18	22	25	29	36
18	02	04	08	12	13	15	17	19	23	27	31	39
19	02	04	08	12	14	16	18	20	24	28	33	41
20	02	04	09	13	15	17	19	21	26	30	34	43
21	02	04	09	13	16	18	20	22	27	31	36	45
22	02	05	09	14	16	19	21	24	28	33	38	47
23	02	05	10	15	17	20	22	25	30	34	39	49
24	03	05	10	15	18	21	23	26	31	36	41	51
25	03	05	11	16	19	21	24	27	32	37	43	53
26	03	06	11	17	19	22	25	28	33	39	44	56
27	03	06	12	17	20	23	26	29	35	40	46	58
28	03	06	12	18	21	24	27	30	36	42	48	60
29	03	06	12	19	22	25	28	31	37	43	50	62
30	03	06	13	19	22	26	29	32	39	45	51	64
33	04	07	14	21	25	28	32	35	42	49	56	70
63	07	13	27	40	47	54	61	67	81	94	1.08	1.35
93	10	20	40	60	70	80	90	99	1.19	1.39	1.59	1.99

78 Dollars.

	½ per ct.	1 per ct.	2 per ct.	3 per ct.	3½ per ct.	4 per ct.	4½ per ct.	5 per ct.	6 per ct.	7 per ct.	8 per ct.	10 per ct.
Years.												
1	39	78	1.56	2.34	2.73	3.12	3.51	3.90	4.68	5.46	6.24	7.80
2	78	1.56	3.12	4.68	5.46	6.24	7.02	7.80	9.36	10.92	12.48	15.60
3	1.17	2.34	4.68	7.02	8.19	9.36	10.53	11.70	14.04	16.38	18.72	23.40
4	1.56	3.12	6.24	9.36	10.92	12.48	14.04	15.60	18.72	21.84	24.96	31.20
5	1.95	3.90	7.80	11.70	13.65	15.60	17.55	19.50	23.40	27.30	31.20	39.00
Months.												
1	03	07	13	20	23	26	29	33	39	46	52	65
2	07	13	26	39	46	52	59	65	78	91	1.04	1.30
3	10	20	39	59	68	78	88	98	1.17	1.37	1.56	1.95
4	13	26	52	78	91	1.04	1.17	1.30	1.56	1.82	2.08	2.60
5	16	33	65	98	1.14	1.30	1.46	1.63	1.95	2.28	2.60	3.25
6	20	39	78	1.17	1.37	1.56	1.76	1.95	2.34	2.73	3.12	3.90
7	23	46	91	1.37	1.59	1.82	2.05	2.28	2.73	3.19	3.64	4.55
8	26	52	1.04	1.56	1.82	2.08	2.34	2.60	3.12	3.64	4.16	5.20
9	29	59	1.17	1.76	2.05	2.34	2.63	2.93	3.51	4.10	4.68	5.85
10	33	65	1.30	1.95	2.28	2.60	2.93	3.25	3.90	4.55	5.20	6.50
11	36	72	1.43	2.15	2.50	2.86	3.22	3.58	4.29	5.01	5.72	7.15
Days.												
1	0	0	0	01	01	01	01	01	01	02	02	02
2	0	0	01	01	02	02	02	02	03	03	03	04
3	0	01	01	02	02	03	03	03	04	05	05	07
4	0	01	02	03	03	03	04	04	05	06	07	09
5	01	01	02	03	04	04	05	05	06	08	09	11
6	01	01	03	04	05	05	06	07	08	09	10	13
7	01	02	03	05	05	06	07	08	09	11	12	15
8	01	02	03	05	06	07	08	09	10	12	14	17
9	01	02	04	06	07	08	09	10	12	14	16	20
10	01	02	04	07	08	09	10	11	13	15	17	22
11	01	02	05	07	08	10	11	12	14	17	19	24
12	01	03	05	08	09	10	12	13	16	18	21	26
13	01	03	06	08	10	11	13	14	17	20	23	28
14	02	03	06	09	11	12	14	15	18	21	24	30
15	02	03	07	10	11	13	15	16	20	23	26	33
16	02	03	07	10	12	14	16	17	21	24	28	35
17	02	04	07	11	13	15	17	18	22	26	29	37
18	02	04	08	12	14	16	18	20	23	27	31	39
19	02	04	08	12	14	16	19	21	25	29	33	41
20	02	04	09	13	15	17	20	22	26	30	35	43
21	02	05	09	14	16	18	20	23	27	32	36	46
22	02	05	10	14	17	19	21	24	29	33	38	48
23	02	05	10	15	17	20	22	25	30	35	40	50
24	03	05	10	16	18	21	23	26	31	36	42	52
25	03	05	11	16	19	22	24	27	33	38	43	54
26	03	06	11	17	20	23	25	28	34	39	45	56
27	03	06	12	18	20	23	26	29	35	41	47	59
28	03	06	12	18	21	24	27	30	36	42	49	61
29	03	06	13	19	22	25	28	31	38	44	50	63
30	03	07	13	20	23	26	29	33	39	46	52	65
33	04	07	14	22	25	29	32	36	43	50	57	72
63	07	14	27	41	48	55	61	68	82	96	1.09	1.37
93	10	20	40	60	71	81	91	1.01	1.21	1.41	1.61	2.02

79 Dollars.

Years.	½ per ct.	1 per ct.	2 per ct.	3 per ct.	3½ per ct.	4 per ct.	4½ per ct.	5 per ct.	6 per ct.	7 per ct.	8 per ct.	10 per ct.
1	40	79	1.58	2.37	2.77	3.16	3.56	3.95	4.74	5.53	6.32	7.90
2	79	1.58	3.16	4.74	5.53	6.32	7.11	7.90	9.48	11.06	12.64	15.80
3	1.19	2.37	4.74	7.11	8.30	9.48	10.67	11.85	14.22	16.59	18.96	23.70
4	1.58	3.16	6.32	9.48	11.06	12.64	14.22	15.80	18.96	22.12	25.28	31.60
5	1.98	3.95	7.90	11.85	13.83	15.80	17.78	19.75	23.70	27.65	31.60	39.50
Months.												
1	03	07	13	20	23	26	30	33	40	46	53	66
2	07	13	26	40	46	53	60	66	79	92	1.05	1.32
3	10	20	40	59	69	79	89	99	1.19	1.38	1.58	1.98
4	13	26	53	79	92	1.05	1.19	1.32	1.58	1.84	2.11	2.63
5	16	33	66	99	1.15	1.32	1.48	1.65	1.98	2.30	2.63	3.29
6	20	40	79	1.19	1.38	1.58	1.78	1.98	2.37	2.77	3.16	3.95
7	23	46	92	1.38	1.61	1.84	2.07	2.30	2.77	3.23	3.69	4.61
8	26	53	1.05	1.58	1.84	2.11	2.37	2.63	3.16	3.69	4.21	5.27
9	30	59	1.19	1.78	2.07	2.37	2.67	2.96	3.56	4.15	4.74	5.93
10	33	66	1.32	1.98	2.30	2.63	2.96	3.29	3.95	4.61	5.27	6.58
11	36	72	1.45	2.17	2.53	2.90	3.26	3.62	4.35	5.07	5.79	7.24
Days.												
1	0	0	0	01	01	01	01	01	01	02	02	02
2	0	0	01	01	02	02	02	02	03	03	04	04
3	0	01	01	02	02	03	03	03	04	05	05	07
4	0	01	02	03	03	04	04	04	05	06	07	09
5	01	01	02	03	04	04	05	05	07	08	09	11
6	01	01	03	04	05	05	06	07	08	09	11	13
7	01	02	03	05	05	06	07	08	09	11	12	15
8	01	02	04	05	06	07	08	09	11	12	14	18
9	01	02	04	06	07	08	09	10	12	14	16	20
10	01	02	04	07	08	09	10	11	13	15	18	22
11	01	02	05	07	08	10	11	12	14	17	19	24
12	01	03	05	08	09	11	12	13	16	18	21	26
13	01	03	06	09	10	11	13	14	17	20	23	29
14	02	03	06	09	11	12	14	15	18	22	25	31
15	02	03	07	10	12	13	15	16	20	23	26	33
16	02	04	07	11	12	14	16	18	21	25	28	35
17	02	04	07	11	13	15	17	19	22	26	30	37
18	02	04	08	12	14	16	18	20	24	28	32	40
19	02	04	08	13	15	17	19	21	25	29	33	42
20	02	04	09	13	15	18	20	22	26	31	35	44
21	02	05	09	14	16	18	21	23	28	32	37	46
22	02	05	10	14	17	19	22	24	29	34	39	48
23	03	05	10	15	18	20	23	25	30	35	40	50
24	03	05	11	16	18	21	24	26	32	37	42	53
25	03	05	11	16	19	22	25	27	33	38	44	55
26	03	06	11	17	20	23	26	29	34	40	46	57
27	03	06	12	18	21	24	27	30	36	41	47	59
28	03	06	12	18	22	25	28	31	37	43	49	61
29	03	06	13	19	22	25	29	32	38	45	51	64
30	03	07	13	20	23	26	30	33	40	46	53	66
33	04	07	14	22	25	29	33	36	43	51	58	72
63	07	14	28	41	48	55	62	69	83	97	1.11	1.38
93	10	20	41	61	71	82	92	1.02	1.22	1.43	1.63	2.04

80 Dollars.

Years.	½ per ct.	1 per ct.	2 per ct.	3 per ct.	3½ per ct.	4 per ct.	4½ per ct.	5 per ct.	6 per ct.	7 per ct.	8 per ct.	10 per ct.
1	40	80	1.60	2.40	2.80	3.20	3.60	4.00	4.80	5.60	6.40	8.00
2	80	1.60	3.20	4.80	5.60	6.40	7.20	8.00	9.60	11.20	12.80	16.00
3	1.20	2.40	4.80	7.20	8.40	9.60	10.80	12.00	14.40	16.80	19.20	24.00
4	1.60	3.20	6.40	9.60	11.20	12.80	14.40	16.00	19.20	22.40	25.60	32.00
5	2.00	4.00	8.00	12.00	14.00	16.00	18.00	20.00	24.00	28.00	32.00	40.00

Months.												
1	03	07	13	20	23	27	30	33	40	47	53	67
2	07	13	27	40	47	53	60	67	80	93	1.07	1.33
3	10	20	40	60	70	80	90	1.00	1.20	1.40	1.60	2.00
4	13	27	53	80	93	1.07	1.20	1.33	1.60	1.87	2.13	2.67
5	17	33	67	1.00	1.17	1.33	1.50	1.67	2.00	2.33	2.67	3.33
6	20	40	80	1.20	1.40	1.60	1.80	2.00	2.40	2.80	3.20	4.00
7	23	47	93	1.40	1.63	1.87	2.10	2.33	2.80	3.27	3.73	4.67
8	27	53	1.07	1.60	1.87	2.13	2.40	2.67	3.20	3.73	4.27	5.33
9	30	60	1.20	1.80	2.10	2.40	2.70	3.00	3.60	4.20	4.80	6.00
10	33	67	1.33	2.00	2.33	2.67	3.00	3.33	4.00	4.67	5.33	6.67
11	37	73	1.47	2.20	2.57	2.93	3.30	3.67	4.40	5.13	5.87	7.33

Days.												
1	0	0	0	01	01	01	01	01	01	02	02	02
2	0	0	01	01	02	02	02	02	03	03	04	04
3	0	01	01	02	02	03	03	03	04	05	05	07
4	0	01	02	03	03	04	04	04	05	06	07	09
5	01	01	02	03	04	04	05	06	07	08	09	11
6	01	01	03	04	05	05	06	07	08	09	11	13
7	01	02	03	05	05	06	07	08	09	11	12	16
8	01	02	04	05	06	07	08	09	11	12	14	18
9	01	02	04	06	07	08	09	10	12	14	16	20
10	01	02	04	07	08	09	10	11	13	16	18	22
11	01	02	05	07	09	10	11	12	15	17	20	24
12	01	03	05	08	09	11	12	13	16	19	21	27
13	01	03	06	09	10	12	13	14	17	20	23	29
14	02	03	06	09	11	12	14	16	19	22	25	31
15	02	03	07	10	12	13	15	17	20	23	27	33
16	02	04	07	11	12	15	17	19	23	26	30	36
17	02	04	08	11	13	15	17	19	23	26	30	38
18	02	04	08	12	14	16	18	20	24	28	32	40
19	02	04	08	13	15	17	19	21	25	30	34	42
20	02	04	09	13	16	18	20	22	27	31	36	44
21	02	05	09	14	16	19	21	23	28	33	37	47
22	02	05	10	15	17	20	22	24	29	34	39	49
23	03	05	10	15	18	20	23	26	31	36	41	51
24	03	05	11	16	19	21	24	27	32	37	43	53
25	03	06	11	17	19	22	25	28	33	39	44	56
26	03	06	12	17	20	23	26	29	35	40	46	58
27	03	06	12	18	21	24	27	30	36	42	48	60
28	03	06	12	19	22	25	28	31	37	44	50	62
29	03	06	13	19	23	26	29	32	39	45	52	64
30	03	07	13	20	23	27	30	33	40	47	53	67
33	04	07	15	22	26	29	33	37	44	51	59	73
63	07	14	28	42	49	56	63	70	84	98	1.12	1.40
93	10	21	41	62	72	83	93	1.03	1.24	1.45	1.65	2.07

81 Dollars.

Years.	½ per ct.	1 per ct.	2 per ct.	3 per ct.	3½ per ct.	4 per ct.	4½ per ct.	5 per ct.	6 per ct.	7 per ct.	8 per ct.	10 per ct
1	41	81	1.62	2.43	2.84	3.24	3.65	4.05	4.86	5.67	6.48	8.10
2	81	1.62	3.24	4.86	5.67	6.48	7.29	8.10	9.72	11.34	12.96	16.20
3	1.22	2.43	4.86	7.29	8.51	9.72	10.94	12.15	14.58	17.01	19.44	24.30
4	1.62	3.24	6.48	9.72	11.34	12.96	14.58	16.20	19.44	22.68	25.92	32.40
5	2.03	4.05	8.10	12.15	14.18	16.20	18.23	20.25	24.30	28.35	32.40	40.50
Months.												
1	03	07	14	20	24	27	30	34	41	47	54	68
2	07	14	27	41	47	54	61	68	81	95	1.08	1.35
3	10	20	41	61	71	81	91	1.01	1.22	1.42	1.62	2.03
4	14	27	54	81	95	1.08	1.22	1.35	1.62	1.89	2.16	2.70
5	17	34	38	1.01	1.18	1.35	1.52	1.69	2.03	2.36	2.70	3.38
6	20	41	81	1.22	1.42	1.62	1.82	2.03	2.43	2.84	3.24	4.05
7	24	47	95	.1.42	1.65	1.89	2.13	2.36	2.84	3.31	3.78	4.73
8	27	54	1.08	1.62	1.89	2.16	2.43	2.70	3.24	3.78	4.32	5.40
9	30	61	1.22	1.82	2.13	2.43	2.73	3.04	3.65	4.25	4.86	6.08
10	34	68	1.35	2.03	2.36	2.70	8.04	3.38	4.05	4.73	5.40	6.75
11	37	74	1.49	2.23	2.60	2.97	3.34	3.71	4.46	5.20	5.94	7.43
Days.												
1	0	0	0	01	01	01	01	01	01	02	02	02
2	0	0	01	01	02	02	02	02	03	03	04	04
3	0	01	01	02	02	03	03	03	04	05	05	07
4	0	01	02	03	03	04	04	04	05	06	07	09
5	01	01	02	03	04	05	05	06	07	08	09	11
6	01	01	03	04	05	05	06	07	08	09	11	14
7	01	02	03	05	06	06	07	08	09	11	13	16
8	01	02	04	05	06	07	08	09	11	13	14	18
9	01	02	04	06	07	08	09	10	12	14	16	20
10	01	02	05	07	08	09	10	11	14	16	18	23
11	01	02	05	07	09	10	11	12	15	17	20	25
12	01	03	05	08	09	11	12	14	16	19	22	27
13	01	03	06	09	10	12	13	15	18	20	23	29
14	02	03	06	09	11	13	14	16	19	22	25	32
15	02	03	07	10	12	14	15	17	20	24	27	34
16	02	04	07	11	13	14	16	18	22	25	29	36
17	02	04	08	11	13	15	17	19	23	27	31	38
18	02	04	08	12	14	16	18	20	24	28	32	41
19	02	04	09	13	15	17	19	21	26	30	34	43
20	02	05	09	14	16	18	20	23	27	32	36	45
21	02	05	09	14	17	19	21	24	28	33	38	47
22	02	05	10	15	17	20	22	25	30	35	40	50
23	03	05	10	16	18	21	23	26	31	36	41	52
24	03	05	11	16	19	22	24	27	32	38	43	54
25	03	06	11	17	20	23	25	28	34	39	45	56
26	03	06	12	18	20	23	26	29	35	41	47	59
27	03	06	12	18	21	24	27	30	36	43	49	61
28	03	06	13	19	22	25	28	32	38	44	50	63
29	03	07	13	20	23	26	29	33	39	46	52	65
30	03	07	14	20	24	27	30	34	41	47	54	68
33	04	07	15	22	26	30	33	37	45	52	59	74
63	07	14	28	43	50	57	64	71	85	99	1.18	1.42
93	10	21	42	63	73	84	94	1.05	1.26	1.46	1.67	2.09

82 Dollars.

	½ per ct.	1 per ct.	2 per ct.	3 per ct.	3½ per ct.	4 per ct.	4½ per ct.	5 per ct.	6 per ct.	7 per ct.	8 per ct.	10 per ct.
Years.												
1	41	82	1.64	2.46	2.87	3.28	3.69	4.10	4.92	5.74	6.56	8.20
2	82	1.64	3.28	4.92	5.74	6.56	7.38	8.20	9.84	11.48	13.12	16.40
3	1.23	2.46	4.92	7.38	8.61	9.84	11.07	12.30	14.76	17.22	19.68	24.60
4	1.64	3.28	6.56	9.84	11.48	13.12	14.76	16.40	19.68	22.96	26.24	32.80
5	2.05	4.10	8.20	12.30	14.35	16.40	18.45	20.50	24.60	28.70	32.80	41.00
Months.												
1	03	07	14	21	24	27	31	34	41	48	55	68
2	07	14	27	41	48	55	62	68	82	96	1.09	1.37
3	10	21	41	62	72	82	92	1.03	1.23	1.44	1.64	2.05
4	14	27	55	82	96	1.09	1.23	1.37	1.64	1.92	2.19	2.73
5	17	34	68	1.03	1.20	1.37	1.54	1.71	2.05	2.39	2.73	3.42
6	21	41	82	1.23	1.44	1.64	1.85	2.05	2.46	2.87	3.28	4.10
7	24	48	96	1.44	1.67	1.91	2.15	2.39	2.87	3.35	3.83	4.78
8	27	55	1.09	1.64	1.91	2.19	2.46	2.73	3.28	3.83	4.37	5.47
9	31	62	1.23	1.85	2.15	2.46	2.77	3.08	3.69	4.31	4.92	6.15
10	34	68	1.37	2.05	2.39	2.73	3.08	3.42	4.10	4.78	5.47	6.83
11	38	75	1.50	2.26	2.63	3.01	3.38	3.76	4.51	5.26	6.01	7.52
Days.												
1	0	0	0	01	01	01	01	01	01	02	02	02
2	0	0	01	01	02	02	02	02	03	03	04	05
3	0	01	01	02	02	03	03	03	04	05	05	07
4	0	01	02	03	03	04	04	05	05	06	07	09
5	01	01	02	03	04	05	05	06	07	08	09	11
6	01	01	03	04	05	05	06	07	08	10	11	14
7	01	02	03	05	06	06	07	08	10	11	13	16
8	01	02	04	05	06	07	08	09	11	13	15	18
9	01	02	04	06	07	08	09	10	12	14	16	21
10	01	02	05	07	08	09	10	11	14	16	18	23
11	01	03	05	08	09	10	11	13	15	18	20	25
12	01	03	05	08	10	11	12	14	16	19	22	27
13	01	03	06	09	10	12	13	15	18	21	24	30
14	02	03	06	10	11	13	14	16	19	22	26	32
15	02	03	07	10	12	14	15	17	21	24	27	34
16	02	04	07	11	13	15	16	18	22	26	29	36
17	02	04	08	12	14	15	17	19	23	27	31	39
18	02	04	08	12	14	16	18	21	25	29	33	41
19	02	04	09	13	15	17	19	22	26	30	35	43
20	02	05	09	14	16	18	21	23	27	32	36	46
21	02	05	10	14	17	19	22	24	29	33	38	48
22	03	05	10	15	18	20	23	25	30	35	40	50
23	03	05	10	16	18	21	24	26	31	37	42	52
24	03	05	11	16	19	22	25	27	33	38	44	55
25	03	06	11	17	20	23	26	28	34	40	46	57
26	03	06	12	18	21	24	27	30	36	41	47	59
27	03	06	12	18	22	25	28	31	37	43	49	62
28	03	06	13	19	22	26	29	32	38	45	51	64
29	03	07	13	20	23	26	30	33	40	46	53	66
30	03	07	14	21	24	27	31	34	41	48	55	68
33	04	08	15	23	26	30	34	38	45	53	60	75
63	07	14	29	43	50	57	65	72	86	1.01	1.15	1.44
93	11	21	42	64	74	85	95	1.06	1.27	1.48	1.69	2.12

83 Dollars.

Years.	½ per ct.	1 per ct.	2 per ct.	3 per ct.	3½ per ct.	4 per ct.	4½ per ct.	5 per ct.	6 per ct.	7 per ct.	8 per ct.	10 per ct.
1	42	83	1.66	2.49	2.91	3.32	3.74	4.15	4.98	5.81	6.64	8.30
2	83	1.66	3.32	4.98	5.81	6.64	7.47	8.30	9.96	11.62	13.28	16.60
3	1.25	2.49	4.98	7.47	8.72	9.96	11.21	12.45	14.94	17.43	19.92	24.90
4	1.66	3.32	6.64	9.96	11.62	13.28	14.94	16.60	19.92	23.24	26.56	33.20
5	2.08	4.15	8.30	12.45	14.53	16.60	18.68	20.75	24.90	29.05	33.20	41.50
Months.												
1	03	07	14	21	24	28	31	35	42	48	55	69
2	07	14	28	42	48	55	62	69	83	97	1.11	1.38
3	10	21	42	62	73	83	93	1.04	1.25	1.45	1.66	2.08
4	14	28	55	83	97	1.11	1.25	1.38	1.66	1.94	2.21	2.77
5	17	35	69	1.04	1.21	1.38	1.56	1.73	2.08	2.42	2.77	3.46
6	21	42	83	1.25	1.45	1.66	1.87	2.08	2.49	2.91	3.32	4.15
7	24	48	97	1.45	1.69	1.94	2.18	2.42	2.91	3.39	3.87	4.84
8	28	55	1.11	1.66	1.94	2.21	2.49	2.77	3.32	3.87	4.43	5.53
9	31	62	1.25	1.87	2.18	2.49	2.80	3.11	3.74	4.36	4.98	6.23
10	35	69	1.38	2.08	2.42	2.77	3.11	3.46	4.15	4.84	5.53	6.92
11	38	76	1.52	2.28	2.66	3.04	3.42	3.80	4.57	5.33	6.09	7.61
Days.												
1	0	0	0	01	01	01	01	01	01	02	02	02
2	0	0	01	01	02	02	02	02	03	03	04	05
3	0	01	01	02	02	03	03	03	04	05	06	07
4	0	01	02	03	03	04	04	05	06	06	07	09
5	01	01	02	03	04	05	05	06	07	08	09	12
6	01	01	03	04	05	06	06	07	08	10	11	14
7	01	02	03	05	06	06	07	08	10	11	13	16
8	01	02	04	06	06	07	08	09	11	13	15	18
9	01	02	04	06	07	08	09	10	12	15	17	21
10	01	02	05	07	08	09	10	12	14	16	18	23
11	01	03	05	08	09	10	11	13	15	18	20	25
12	01	03	06	08	10	11	12	14	17	19	22	28
13	01	03	06	09	10	12	13	15	18	21	24	30
14	02	03	06	10	11	13	15	16	19	23	26	32
15	02	03	07	10	12	14	16	17	21	24	28	35
16	02	04	07	11	13	15	17	18	22	26	30	37
17	02	04	08	12	14	16	18	20	24	27	31	39
18	02	04	08	12	15	17	19	21	25	29	33	42
19	02	04	09	13	15	18	20	22	26	31	35	44
20	02	05	09	14	16	18	21	23	28	32	37	46
21	02	05	10	14	17	19	22	24	29	34	39	48
22	03	05	10	15	18	20	23	25	30	36	41	51
23	03	05	11	16	19	21	24	27	32	37	42	53
24	03	06	11	17	19	22	25	28	33	39	44	55
25	03	06	12	17	20	23	26	29	35	40	46	58
26	03	06	12	18	21	24	27	30	36	42	48	60
27	03	06	12	19	22	25	28	31	37	44	50	62
28	03	06	13	19	23	26	29	32	39	45	52	65
29	03	07	13	20	23	27	30	33	40	47	53	67
30	03	07	14	21	24	28	31	35	42	48	55	69
33	04	08	15	23	27	30	34	38	46	53	61	76
63	07	15	29	44	51	58	65	73	87	1.02	1.16	1.45
93	11	21	43	64	75	86	96	1.07	1.29	1.50	1.72	2.14

84 Dollars.

	½ per ct.	1 per ct.	2 per ct.	3 per ct.	3½ per ct.	4 per ct.	4½ per ct.	5 per ct.	6 per ct.	7 per ct.	8 per ct.	10 per ct.
Years.												
1	42	84	1.68	2.52	2.94	3.36	3.78	4.20	5.04	5.88	6.72	8.40
2	84	1.68	3.36	5.04	5.88	6.72	7.56	8.40	10.08	11.76	13.44	16.80
3	1.26	2.52	5.04	7.56	8.82	10.08	11.34	12.60	15.12	17.64	20.16	25.20
4	1.68	3.36	6.72	10.08	11.76	13.44	15.12	16.80	20.16	23.52	26.88	33.60
5	2.10	4.20	8.40	12.60	14.70	16.80	18.90	21.00	25.20	29.40	33.60	42.00
Months.												
1	04	07	14	21	25	28	32	35	42	49	56	70
2	07	14	28	42	49	56	63	70	84	98	1.12	1.40
3	11	21	42	63	74	84	95	1.05	1.26	1.47	1.68	2.10
4	14	28	56	84	98	1.12	1.26	1.40	1.68	1.96	2.24	2.80
5	18	35	70	1.05	1.23	1.40	1.58	1.75	2.10	2.45	2.80	3.50
6	21	42	84	1.26	1.47	1.68	1.89	2.10	2.52	2.94	3.36	4.20
7	25	49	98	1.47	1.72	1.96	2.21	2.45	2.94	3.43	3.92	4.90
8	28	56	1.12	1.68	1.96	2.24	2.52	2.80	3.36	3.92	4.48	5.60
9	32	63	1.26	1.89	2.21	2.52	2.84	3.15	3.78	4.41	5.04	6.30
10	35	70	1.40	2.10	2.45	2.80	3.15	3.50	4.20	4.90	5.60	7.00
11	39	77	1.54	2.31	2.70	3.08	3.47	3.85	4.62	5.39	6.16	7.70
Days.												
1	0	0	0	01	01	01	01	01	01	02	02	02
2	0	0	01	01	02	02	02	02	03	03	04	05
3	0	01	01	02	02	03	03	03	04	05	06	07
4	0	01	02	03	03	04	04	05	06	07	07	09
5	01	01	02	04	04	05	05	06	07	08	09	12
6	01	01	03	04	05	06	06	07	08	10	11	14
7	01	02	03	05	06	07	07	08	10	11	13	16
8	01	02	04	06	07	08	08	09	11	13	15	19
9	01	02	04	06	07	08	09	11	13	15	17	21
10	01	02	05	07	08	09	11	12	14	16	19	23
11	01	03	05	08	09	10	12	13	15	18	21	26
12	01	03	06	08	10	11	13	14	17	20	22	28
13	01	03	06	09	11	12	14	15	18	21	24	30
14	02	03	07	10	11	13	15	16	20	23	26	33
15	02	04	07	11	12	14	16	18	21	25	28	35
16	02	04	07	11	13	15	17	19	22	26	30	37
17	02	04	08	12	14	16	18	20	24	28	32	40
18	02	04	08	13	15	17	19	21	25	29	34	42
19	02	04	09	13	16	18	20	22	27	31	35	44
20	02	05	09	14	16	19	21	23	28	33	37	47
21	02	05	10	15	17	20	22	25	29	34	39	49
22	03	05	10	15	18	21	23	26	31	36	41	51
23	03	05	11	16	19	21	24	27	32	38	43	54
24	03	06	11	17	20	22	25	28	34	39	45	56
25	03	06	12	18	20	23	26	29	35	41	47	58
26	03	06	12	18	21	24	27	30	36	42	49	61
27	03	06	13	19	22	25	28	32	38	44	50	63
28	03	07	13	20	23	26	29	33	39	46	52	65
29	03	07	14	20	24	27	30	34	41	47	54	68
30	04	07	14	21	25	28	32	35	42	49	56	70
33	04	08	15	23	27	31	35	39	46	54	62	77
63	07	15	29	44	51	59	66	74	88	1.03	1.18	1.47
93	11	22	43	65	76	87	98	1.09	1.30	1.52	1.74	2.17

85 Dollars.

Years.	¼ per ct.	1 per ct.	2 per ct.	3 per ct.	3½ per ct.	4 per ct.	4½ per ct.	5 per ct.	6 per ct.	7 per ct.	8 per ct.	10 per ct.
1	43	85	1.70	2.55	2.98	3.40	3.83	4.25	5.10	5.95	6.80	8.50
2	85	1.70	3.40	5.10	5.95	6.80	7.65	8.50	10.20	11.90	13.60	17.00
3	1.28	2.55	5.10	7.65	8.93	10.20	11.48	12.75	15.30	17.85	20.40	25.50
4	1.70	3.40	6.80	10.20	11.90	13.60	15.30	17.00	20.40	23.80	27.20	34.00
5	2.13	4.25	8.50	12.75	14.88	17.00	19.13	21.25	25.50	29.75	34.00	42.50

Months.												
1	04	07	14	21	25	28	32	35	43	50	57	71
2	07	14	28	43	50	57	64	71	85	99	1.13	1.42
3	11	21	43	64	74	85	96	1.06	1.28	1.49	1.70	2.13
4	14	28	57	85	99	1.13	1.28	1.42	1.70	1.98	2.27	2.83
5	18	35	71	1.06	1.24	1.42	1.59	1.77	2.13	2.48	2.83	3.54
6	21	43	85	1.28	1.49	1.70	1.91	2.13	2.55	2.98	3.40	4.25
7	25	50	99	1.49	1.74	1.98	2.23	2.48	2.98	3.47	3.97	4.96
8	28	57	1.13	1.70	1.98	2.27	2.55	2.83	3.40	3.97	4.53	5.67
9	32	64	1.28	1.91	2.23	2.55	2.87	3.19	3.83	4.46	5.10	6.38
10	35	71	1.42	2.13	2.48	2.83	3.19	3.54	4.25	4.96	5.67	7.08
11	39	78	1.56	2.34	2.73	3.12	3.51	3.90	4.68	5.45	6.23	7.79

Days.												
1	0	0	0	01	01	01	01	01	01	02	02	02
2	0	0	01	01	02	02	02	02	03	03	04	05
3	0	01	01	02	02	03	03	04	04	05	06	07
4	0	01	02	03	03	04	04	05	06	07	08	09
5	01	01	02	04	04	05	05	06	07	08	09	12
6	01	01	03	04	05	06	06	07	09	10	11	14
7	01	02	03	05	06	07	07	08	10	12	13	17
8	01	02	04	06	07	08	08	09	11	13	15	19
9	01	02	04	06	07	09	10	11	13	15	17	21
10	01	02	05	07	08	09	11	12	14	17	19	24
11	01	03	05	08	09	10	12	13	16	18	21	26
12	01	03	06	09	16	11	13	14	17	20	23	28
13	02	03	06	09	11	12	14	15	18	21	25	31
14	02	03	07	10	12	13	15	17	20	23	26	33
15	02	04	07	11	12	14	16	18	21	25	28	35
16	02	04	08	11	13	15	17	19	23	26	30	38
17	02	04	08	12	14	16	18	20	24	28	32	40
18	02	04	09	13	15	17	19	21	26	30	34	43
19	02	04	09	13	16	18	20	22	27	31	36	45
20	02	05	09	14	17	19	21	24	28	33	38	47
21	02	05	10	15	17	20	22	25	30	35	40	50
22	03	05	10	16	18	21	23	26	31	36	42	52
23	03	05	11	16	19	22	24	27	33	38	43	54
24	03	06	11	17	20	23	26	28	34	40	45	57
25	03	06	12	18	21	24	27	30	35	41	47	59
26	03	06	12	18	21	25	28	31	37	43	49	61
27	03	06	13	19	22	26	29	32	38	45	51	64
28	03	07	13	20	23	26	30	33	40	46	53	66
29	03	07	14	21	24	27	31	34	41	48	55	68
30	04	07	14	21	25	28	32	35	43	50	57	71

33	04	08	16	23	27	31	35	39	47	55	62	78
63	07	15	30	45	52	60	67	74	89	1.04	1.19	1.49
93	11	22	44	66	77	88	99	1.10	1.32	1.54	1.76	2.20

86 Dollars.

	¼ per ct.	1 per ct.	2 per ct.	3 per ct.	3½ per ct.	4 per ct.	4½ per ct.	5 per ct.	6 per ct.	7 per ct.	8 per ct.	10 per ct.
Years.												
1	43	86	1.72	2.58	3.01	3.44	3.87	4.30	5.16	6.02	6.88	8.60
2	86	1.72	3.44	5.16	6.02	6.88	7.74	8.60	10.32	12.04	13.76	17.20
3	1.29	2.58	5.16	7.74	9.03	10.32	11.61	12.90	15.48	18.06	20.64	25.80
4	1.72	3.44	6.88	10.32	12.04	13.76	15.48	17.20	20.64	24.08	27.52	34.40
5	2.15	4.30	8.60	12.90	15.05	17.20	19.35	21.50	25.80	30.10	34.40	43.00
Months.												
1	04	07	14	22	25	29	32	36	43	50	57	72
2	07	14	29	43	50	57	65	72	86	1.00	1.15	1.43
3	11	22	43	65	75	86	97	1.08	1.29	1.51	1.72	2.15
4	14	29	57	86	1.00	1.15	1.29	1.43	1.72	2.01	2.29	2.87
5	18	36	72	1.08	1.25	1.43	1.61	1.79	2.15	2.51	2.87	3.58
6	22	43	86	1.29	1.51	1.72	1.94	2.15	2.58	3.01	3.44	4.30
7	25	50	1.00	1.51	1.76	2.01	2.26	2.51	3.01	3.51	4.01	5.02
8	29	57	1.15	1.72	2.01	2.29	2.58	2.87	3.44	4.01	4.59	5.73
9	32	65	1.29	1.94	2.26	2.58	2.90	8.23	3.87	4.52	5.16	6.45
10	36	72	1.43	2.15	2.51	2.87	3.23	3.58	4.30	5.02	5.73	7.17
11	39	79	1.58	2.37	2.76	3.15	3.55	3.94	4.73	5.52	6.31	7.88
Days.												
1	0	0	0	01	01	01	01	01	01	02	02	02
2	0	0	01	01	02	02	02	02	03	03	04	05
3	0	01	01	02	03	03	03	04	04	05	06	07
4	0	01	02	03	03	04	04	05	06	07	08	10
5	01	01	02	04	04	05	05	06	07	08	10	12
6	01	01	03	04	05	06	06	07	09	10	11	14
7	01	02	03	05	06	07	08	08	10	12	13	17
8	01	02	04	06	07	08	09	10	11	13	15	19
9	01	02	04	06	08	09	10	11	13	15	17	22
10	01	02	05	07	08	10	11	12	14	17	19	24
11	01	03	05	08	09	11	12	13	16	18	21	26
12	01	03	06	09	10	11	13	14	17	20	23	29
13	02	03	06	09	11	12	14	16	19	22	25	31
14	02	03	07	10	12	13	15	17	20	23	27	33
15	02	04	07	11	13	14	16	18	22	25	29	36
16	02	04	08	11	13	15	17	19	23	27	31	38
17	02	04	08	12	14	16	18	20	24	28	32	41
18	02	04	09	13	15	17	19	22	26	30	34	43
19	02	05	09	14	16	18	20	23	27	32	36	45
20	02	05	10	14	17	19	22	24	29	33	38	48
21	03	05	10	15	18	20	23	25	30	35	40	50
22	03	05	11	16	18	21	24	26	32	37	42	53
23	03	05	11	16	19	22	25	27	33	38	44	55
24	03	06	11	17	20	23	26	29	34	40	46	57
25	03	06	12	18	21	24	27	30	36	42	48	60
26	03	06	12	19	22	25	28	31	37	43	50	62
27	03	06	13	19	23	26	29	32	39	45	52	65
28	03	07	13	20	23	27	30	33	40	47	54	67
29	03	07	14	21	24	28	31	35	42	48	55	69
30	04	07	14	22	25	29	32	36	43	50	57	72
33	04	08	16	24	28	32	35	39	47	55	63	79
63	08	15	30	45	53	60	68	76	90	1.05	1.20	1.51
93	11	22	44	67	78	89	1.00	1.11	1.33	1.56	1.78	2.22

87 Dollars.

Years.	½ per ct.	1 per ct.	2 per ct.	3 per ct.	3½ per ct.	4 per ct.	4½ per ct.	5 per ct.	6 per ct.	7 per ct.	8 per ct.	10 per ct.
1	44	87	1.74	2.61	3.05	3.48	3.92	4.35	5.22	6.09	6.96	8.70
2	87	1.74	3.48	5.22	6.09	6.96	7.83	8.70	10.44	12.18	13.92	17.40
3	1.31	2.61	5.22	7.83	9.14	10.44	11.75	13.05	15.66	18.27	20.88	26.10
4	1.74	3.48	6.96	10.44	12.18	13.92	15.66	17.40	20.88	24.36	27.84	34.80
5	2.18	4.35	8.70	13.05	15.23	17.40	19.58	21.75	26.10	30.45	34.80	43.50
Months.												
1	04	07	15	22	25	29	33	36	44	51	58	73
2	07	15	29	44	51	58	65	73	87	1.02	1.16	1.45
3	11	22	44	65	76	87	98	1.09	1.31	1.52	1.74	2.18
4	15	29	58	87	1.02	1.16	1.31	1.45	1.74	2.03	2.32	2.90
5	18	36	73	1.09	1.27	1.45	1.63	1.81	2.18	2.54	2.90	3.63
6	22	44	87	1.31	1.52	1.74	1.96	2.18	2.61	3.05	3.48	4.35
7	25	51	1.02	1.52	1.78	2.03	2.28	2.54	3.05	3.55	4.06	5.08
8	29	58	1.16	1.74	2.03	2.32	2.61	2.90	3.48	4.06	4.64	5.80
9	33	65	1.31	1.96	2.28	2.61	2.94	3.26	3.92	4.57	5.22	6.53
10	36	73	1.45	2.18	2.54	2.90	3.26	3.63	4.35	5.08	5.80	7.25
11	40	80	1.60	2.39	2.79	3.19	3.59	3.99	4.79	5.58	6.38	7.98
Days.												
1	0	0	0	01	01	01	01	01	01	02	02	02
2	0	0	01	01	02	02	02	02	03	03	04	05
3	0	01	01	02	03	03	03	04	04	05	06	07
4	0	01	02	03	03	04	04	05	06	07	08	10
5	01	01	02	04	04	05	05	06	07	08	10	12
6	01	01	03	04	05	06	07	07	09	10	12	15
7	01	02	03	05	06	07	08	08	10	12	14	17
8	01	02	04	06	07	08	09	10	12	14	15	19
9	01	02	04	07	08	09	10	11	13	15	17	22
10	01	02	05	07	08	10	11	12	15	17	19	24
11	01	03	05	08	09	11	12	13	16	19	21	27
12	01	03	06	09	10	12	13	15	17	20	23	29
13	02	03	06	09	11	13	14	16	19	22	25	31
14	02	03	07	10	12	14	15	17	20	24	27	34
15	02	04	07	11	13	15	16	18	22	25	29	36
16	02	04	08	12	14	15	17	19	23	27	31	39
17	02	04	08	12	14	16	18	21	25	29	33	41
18	02	04	09	13	15	17	20	22	26	30	35	44
19	02	05	09	14	16	18	21	23	28	32	37	46
20	02	05	10	15	17	19	22	24	29	34	39	48
21	03	05	10	15	18	20	23	25	30	36	41	51
22	03	05	11	16	19	21	24	27	32	37	43	53
23	03	06	11	17	19	22	25	28	33	39	44	56
24	03	06	12	17	20	23	26	29	35	41	46	58
25	03	06	12	18	21	24	27	30	36	42	48	60
26	03	06	13	19	22	25	28	31	38	44	50	63
27	03	07	13	20	23	26	29	33	39	46	52	65
28	03	07	14	20	24	27	30	34	41	47	54	68
29	04	07	14	21	25	28	32	35	42	49	56	70
30	04	07	15	22	25	29	33	36	44	51	58	73
33	04	08	16	24	28	32	36	40	48	56	64	80
63	08	15	30	46	53	61	69	76	91	1.07	1.22	1.52
93	11	22	45	67	79	90	1.01	1.12	1.35	1.57	1.80	2.25

88 Dollars.

	$\frac{1}{2}$ per ct.	1 per ct.	2 per ct.	3 per ct.	$3\frac{1}{2}$ per ct.	4 per ct.	$4\frac{1}{2}$ per ct.	5 per ct.	6 per ct.	7 per ct.	8 per ct.	10 per ct.
Years.												
1	44	88	1.76	2.64	3.08	3.52	3.96	4.40	5.28	6.16	7.04	8.80
2	88	1.76	3.52	5.28	6.16	7.04	7.92	8.80	10.56	12.32	14.08	17.60
3	1.32	2.64	5.28	7.92	9.24	10.56	11.88	13.20	15.84	18.48	21.12	26.40
4	1.76	3.52	7.04	10.56	12.32	14.08	15.84	17.60	21.12	24.64	28.16	35.20
5	2.20	4.40	8.80	13.20	15.40	17.60	19.80	22.00	26.40	30.80	35.20	44.00
Months.												
1	04	07	15	22	26	29	33	37	44	51	59	73
2	07	15	29	44	51	59	66	73	88	1.03	1.17	1.47
3	11	22	44	66	77	88	99	1.10	1.32	1.54	1.76	2.20
4	15	29	59	88	1.03	1.17	1.32	1.47	1.76	2.05	2.35	2.93
5	18	37	73	1.10	1.28	1.47	1.65	1.83	2.20	2.57	2.93	3.67
6	22	44	88	1.32	1.54	1.76	1.98	2.20	2.64	3.08	3.52	4.40
7	26	51	1.03	1.54	1.80	2.05	2.31	2.57	3.08	3.59	4.11	5.13
8	29	59	1.17	1.76	2.05	2.35	2.64	2.93	3.52	4.11	4.69	5.87
9	33	66	1.32	1.98	2.31	2.64	2.97	3.30	3.96	4.62	5.28	6.60
10	37	73	1.47	2.20	2.57	2.93	3.30	3.67	4.40	5.13	5.87	7.33
11	40	81	1.61	2.42	2.82	3.23	3.63	4.03	4.84	5.65	6.45	8.07
Days.												
1	0	0	0	01	01	01	01	01	01	02	02	02
2	0	0	01	01	02	02	02	02	03	03	04	05
3	0	01	01	02	03	03	03	04	04	05	06	07
4	0	01	02	03	03	04	04	05	06	07	08	10
5	01	01	02	04	04	05	06	06	07	09	10	12
6	01	01	03	04	05	06	07	07	09	10	12	15
7	01	02	03	05	06	07	08	09	10	12	14	17
8	01	02	04	06	07	08	09	10	12	14	16	20
9	01	02	04	07	08	09	10	11	13	15	18	22
10	01	02	05	07	09	10	11	12	15	17	20	24
11	01	03	05	08	09	11	12	13	16	19	22	27
12	01	03	06	09	10	12	13	15	18	21	23	29
13	02	03	06	10	11	13	14	16	19	22	25	32
14	02	03	07	10	12	14	15	17	21	24	27	34
15	02	04	07	11	13	15	17	18	22	26	29	37
16	02	04	08	12	14	16	18	20	23	27	31	39
17	02	04	08	12	15	17	19	21	25	29	33	42
18	02	04	09	13	15	18	20	22	26	31	35	44
19	02	05	09	14	16	19	21	23	28	33	37	46
20	02	05	10	15	17	20	22	24	29	34	39	49
21	03	05	10	15	18	21	23	26	31	36	41	51
22	03	05	11	16	19	22	24	27	32	38	43	54
23	03	06	11	17	20	22	25	28	34	39	45	56
24	03	06	12	18	21	23	26	29	35	41	47	59
25	03	06	12	18	21	24	28	31	37	43	49	61
26	03	06	13	19	22	25	29	32	38	44	51	64
27	03	07	13	20	23	26	30	33	40	46	53	66
28	03	07	14	21	24	27	31	34	41	48	55	68
29	04	07	14	21	25	28	32	35	43	50	57	71
30	04	07	15	22	26	29	33	37	44	51	59	73
33	04	08	16	24	28	32	36	40	48	56	65	81
63	08	15	31	46	54	62	69	77	92	1.08	1.23	1.54
93	11	23	45	68	80	91	1.02	1.14	1.36	1.59	1.82	2.27

89 Dollars.

Years.	½ per ct.	1 per ct.	2 per ct.	3 per ct.	3½ per ct.	4 per ct.	4½ per ct.	5 per ct.	6 per ct.	7 per ct.	8 per ct.	10 per ct.
1	45	89	1.78	2.67	3.12	3.56	4.01	4.45	5.34	6.23	7.12	8.90
2	89	1.78	3.56	5.34	6.23	7.12	8.01	8.90	10.68	12.46	14.24	17.80
3	1.34	2.67	5.34	8.01	9.35	10.68	12.02	13.35	16.02	18.69	21.36	26.70
4	1.78	3.56	7.12	10.68	12.46	14.24	16.02	17.80	21.36	24.92	28.48	35.60
5	2.26	4.45	8.90	13.35	15.58	17.80	20.03	22.25	26.70	31.15	35.60	44.50
Months.												
1	04	07	15	22	26	30	33	37	45	52	59	74
2	08	15	30	45	52	59	67	74	89	1.04	1.19	1.48
3	11	22	45	67	78	89	1.00	1.11	1.34	1.56	1.78	2.23
4	15	33	59	89	1.04	1.19	1.34	1.48	1.78	2.08	2.37	2.97
5	19	37	74	1.11	1.30	1.48	1.67	1.85	2.23	2.60	2.97	3.71
6	22	45	89	1.34	1.56	1.78	2.00	2.23	2.67	3.12	3.56	4.45
7	26	52	1.04	1.56	1.82	2.08	2.34	2.60	3.12	3.63	4.15	5.19
8	30	59	1.19	1.78	2.08	2.37	2.67	2.97	3.56	4.15	4.75	5.93
9	53	67	1.34	2.00	2.34	2.67	3.00	3.34	4.01	4.67	5.34	6.68
10	37	74	1.48	2.23	2.60	2.97	3.34	3.71	4.45	5.19	5.93	7.42
11	41	82	1.63	2.45	2.86	3.26	3.67	4.08	4.90	5.71	6.53	8.16
Days.												
1	0	0	0	01	01	01	01	01	01	02	02	02
2	0	0	01	01	02	02	02	02	03	03	04	05
3	0	01	01	02	03	03	03	04	04	05	06	07
4	0	01	02	03	03	04	04	05	06	07	08	10
5	01	01	02	04	04	05	06	06	07	09	10	12
6	01	01	03	04	05	06	07	07	09	10	12	15
7	01	02	03	05	06	07	08	09	10	12	14	17
8	01	02	04	06	07	08	09	10	12	14	16	20
9	01	02	04	07	08	09	10	11	13	16	18	22
10	01	02	05	07	09	10	11	12	15	17	20	25
11	01	03	05	08	10	11	12	14	16	19	22	27
12	01	03	06	09	10	12	13	15	18	21	24	30
13	02	03	06	10	11	13	14	16	19	22	26	32
14	02	03	07	10	12	14	16	17	21	24	28	35
15	02	04	07	11	13	15	17	19	22	26	30	37
16	02	04	08	12	14	16	18	20	24	28	32	40
17	02	04	08	13	15	17	19	21	25	29	34	42
18	02	04	09	13	16	18	20	22	27	31	36	45
19	02	05	09	14	16	19	21	23	28	33	38	47
20	02	05	10	15	17	20	22	25	30	35	40	49
21	03	05	10	16	18	21	23	26	31	36	42	52
22	03	05	11	16	19	22	24	27	33	38	44	54
23	03	06	11	17	20	23	26	28	34	40	45	57
24	03	06	12	18	21	24	27	30	36	42	47	59
25	03	06	12	19	22	25	28	31	37	43	49	62
26	03	06	13	19	22	26	29	32	39	45	51	64
27	03	07	13	20	23	27	30	33	40	47	53	67
28	03	07	14	21	24	28	31	35	42	48	55	69
29	04	07	14	22	25	29	32	36	43	50	57	72
30	04	07	15	22	26	30	33	37	45	52	59	74
33	04	08	16	24	29	33	37	41	49	57	65	82
63	08	16	31	47	55	62	70	78	93	1.09	1.25	1.56
93	11	23	46	69	80	92	1.03	1.15	1.38	1.61	1.84	2.30

90 Dollars.

Years.	½ per ct.	1 per ct.	2 per ct.	3 per ct.	3½ per ct.	4 per ct.	4½ per ct.	5 per ct.	6 per ct.	7 per ct.	8 per ct.	10 per ct.
1	45	90	1.80	2.70	3.15	3.60	4.05	4.50	5.40	6.30	7.20	9.00
2	90	1.80	3.60	5.40	6.30	7.20	8.10	9.00	10.80	12.60	14.40	18.00
3	1.35	2.70	5.40	8.10	9.45	10.80	12.15	13.50	16.20	18.90	21.60	27.00
4	1.80	3.60	7.20	10.80	12.60	14.40	16.20	18.00	21.60	25.20	28.80	36.00
5	2.25	4.50	9.00	13.50	15.75	18.00	20.25	22.50	27.00	31.50	36.00	45.00
Months.												
1	04	08	15	23	26	30	34	38	45	53	60	75
2	08	15	30	45	53	60	68	75	90	1.05	1.20	1.50
3	11	23	45	68	79	90	1.01	1.13	1.35	1.58	1.80	2.25
4	15	30	60	90	1.05	1.20	1.35	1.50	1.80	2.10	2.40	3.00
5	19	38	75	1.13	1.31	1.50	1.69	1.88	2.25	2.63	3.00	3.75
6	23	45	90	1.35	1.58	1.80	2.03	2.25	2.70	3.15	3.60	4.50
7	26	53	1.05	1.58	1.84	2.10	2.36	2.63	3.15	3.68	4.20	5.25
8	30	60	1.20	1.80	2.10	2.40	2.70	3.00	3.60	4.20	4.80	6.00
9	34	68	1.35	2.03	2.36	2.70	3.04	3.38	4.05	4.73	5.40	6.75
10	38	75	1.50	2.25	2.63	3.00	3.38	3.75	4.50	5.25	6.00	7.50
11	41	83	1.65	2.48	2.89	3.30	3.71	4.13	4.95	5.78	6.60	8.25
Days.												
1	0	0	01	01	01	01	01	01	02	02	02	03
2	0	01	01	01	02	02	02	03	03	04	04	05
3	0	01	02	02	03	03	03	04	05	05	06	08
4	01	01	02	03	04	04	04	05	06	07	08	10
5	01	01	03	04	04	05	06	06	08	09	10	13
6	01	02	03	05	05	06	07	08	09	11	12	15
7	01	02	04	05	06	07	08	09	11	12	14	18
8	01	02	04	06	07	08	09	10	12	14	16	20
9	01	02	05	07	08	09	10	11	14	16	18	23
10	01	03	05	08	09	10	11	13	15	18	20	25
11	01	03	06	08	10	11	12	14	17	19	22	28
12	02	03	06	09	11	12	14	15	18	21	24	30
13	02	03	07	10	11	13	15	16	20	23	26	33
14	02	04	07	11	12	14	16	18	21	25	28	35
15	02	04	08	11	13	15	17	19	23	26	30	38
16	02	04	08	12	14	16	18	20	24	28	32	40
17	02	04	09	13	15	17	19	21	26	30	34	43
18	02	05	09	14	16	18	20	23	27	32	36	45
19	02	05	10	14	17	19	21	24	29	33	38	48
20	03	05	10	15	18	20	23	25	30	35	40	50
21	03	05	11	16	18	21	24	26	32	37	42	53
22	03	06	11	17	19	22	25	28	33	39	44	55
23	03	06	12	17	20	23	26	29	35	40	46	58
24	03	06	12	18	21	24	27	30	36	42	48	60
25	03	06	13	19	22	25	28	31	38	44	50	63
26	03	07	13	20	23	26	29	33	39	46	52	65
27	03	07	14	20	24	27	30	34	41	47	54	68
28	04	07	14	21	25	28	32	35	42	49	56	70
29	04	07	15	22	25	29	33	36	44	51	58	73
30	04	08	15	23	26	30	34	38	45	53	60	75
33	04	08	17	25	29	33	37	41	50	58	66	83
63	08	16	32	47	55	63	71	79	95	1.10	1.26	1.58
93	12	23	47	70	81	93	1.05	1.16	1.40	1.63	1.86	2.33

91 Dollars.

Years.	½ per ct.	1 per ct.	2 per ct.	3 per ct.	3½ per ct.	4 per ct.	4½ per ct.	5 per ct.	6 per ct.	7 per ct.	8 per ct.	10 per ct.
1	46	91	1.82	2.73	3.18	3.64	4.10	4.55	5.46	6.37	7.28	9.10
2	91	1.82	3.64	5.46	6.35	7.28	8.19	9.10	10.92	12.74	14.56	18.20
3	1.37	2.73	5.46	8.19	9.53	10.92	12.29	13.65	16.38	19.11	21.84	27.30
4	1.82	3.64	7.28	10.92	12.70	14.56	16.38	18.20	21.84	25.48	29.12	36.40
5	2.28	4.55	9.10	13.65	15.86	18.20	20.48	22.75	27.30	31.85	36.40	45.50
Months.												
1	04	08	15	23	27	30	34	38	46	53	61	76
2	08	15	30	46	53	61	68	76	91	1.06	1.21	1.52
3	11	23	46	68	80	91	1.02	1.14	1.37	1.59	1.82	2.28
4	15	30	61	91	1.06	1.21	1.37	1.52	1.82	2.12	2.43	3.03
5	19	38	76	1.14	1.33	1.52	1.71	1.90	2.28	2.65	3.03	3.79
6	23	46	91	1.37	1.59	1.82	2.05	2.28	2.73	3.19	3.64	4.55
7	27	53	1.06	1.59	1.86	2.12	2.39	2.65	3.19	3.72	4.25	5.31
8	30	61	1.21	1.82	2.12	2.43	2.73	3.03	3.64	4.25	4.85	6.07
9	34	68	1.37	2.05	2.39	2.73	3.07	3.41	4.10	4.78	5.46	6.83
10	38	76	1.52	2.28	2.65	3.03	3.41	3.79	4.55	5.31	6.07	7.58
11	42	83	1.67	2.50	2.92	3.34	3.75	4.17	5.01	5.84	6.67	8.34
Days.												
1	0	0	01	01	01	01	01	01	02	02	02	03
2	0	0	01	02	02	02	02	03	03	04	04	05
3	0	01	02	02	03	03	03	04	05	05	06	08
4	01	01	02	03	04	04	05	05	06	07	08	10
5	01	01	03	04	04	05	05	06	08	09	10	13
6	01	02	03	05	05	06	07	08	09	11	12	15
7	01	02	04	05	06	07	08	09	11	12	14	18
8	01	02	04	06	07	08	09	10	12	14	16	20
9	01	02	05	07	08	09	10	11	14	16	18	23
10	01	03	05	08	09	10	11	13	15	18	20	25
11	01	03	06	08	10	11	13	14	17	19	22	28
12	02	03	06	09	11	12	14	15	18	21	24	30
13	02	03	07	10	12	13	15	16	20	23	26	33
14	02	04	07	11	12	14	16	18	21	25	28	35
15	02	04	08	11	13	15	17	19	23	27	30	38
16	02	04	08	12	14	16	18	20	24	28	32	40
17	02	04	09	13	15	17	19	21	26	30	34	43
18	02	05	09	14	16	18	20	23	27	32	36	46
19	02	05	10	14	17	19	22	24	29	34	38	48
20	03	05	10	15	18	20	23	25	30	35	40	51
21	03	05	11	16	19	21	24	27	32	37	42	53
22	03	06	11	17	19	22	25	28	33	39	44	56
23	03	06	12	17	20	23	26	29	35	41	47	58
24	03	06	12	18	21	24	27	30	36	42	49	61
25	03	06	13	19	22	25	28	32	38	44	51	63
26	03	07	13	20	23	26	30	33	39	46	53	66
27	03	07	14	20	24	27	31	34	41	48	55	68
28	04	07	14	21	25	28	32	35	42	50	57	71
29	04	07	15	22	26	29	33	37	44	51	59	73
30	04	08	15	23	27	30	34	38	46	53	61	76
33	04	08	17	25	29	33	38	42	50	58	67	83
63	08	16	32	48	56	64	72	80	96	1.11	1.27	1.59
93	12	24	47	71	82	94	1.06	1.18	1.41	1.65	1.88	2.35

92 Dollars.

	½ per ct.	1 per ct.	2 per ct.	3 per ct.	3½ per ct.	4 per ct.	4½ per ct.	5 per ct.	6 per ct.	7 per ct.	8 per ct.	10 per ct.
Years.												
1	46	92	1.84	2.76	3.22	3.68	4.14	4.60	5.52	6.44	7.36	9.20
2	92	1.84	3.68	5.52	6.44	7.36	8.28	9.20	11.04	12.88	14.72	18.40
3	1.38	2.76	5.52	8.28	9.66	11.04	12.42	13.80	16.56	19.32	22.08	27.60
4	1.84	3.68	7.36	11.04	12.88	14.72	16.56	18.40	22.08	25.76	29.44	36.80
5	2.30	4.60	9.20	13.80	16.10	18.40	20.70	23.00	27.60	32.20	36.80	46.00
Months.												
1	04	08	15	23	27	31	35	38	46	54	61	77
2	08	15	31	46	54	61	69	77	92	1.07	1.23	1.53
3	12	23	46	69	81	92	1.04	1.15	1.38	1.61	1.84	2.30
4	15	31	61	92	1.07	1.23	1.38	1.53	1.84	2.15	2.45	3.07
5	19	38	77	1.15	1.34	1.53	1.73	1.92	2.30	2.68	3.07	3.83
6	23	46	92	1.38	1.61	1.84	2.07	2.30	2.76	3.22	3.68	4.60
7	27	54	1.07	1.61	1.88	2.15	2.42	2.68	3.22	3.76	4.29	5.37
8	31	61	1.23	1.84	2.15	2.45	2.76	3.07	3.68	4.29	4.91	6.13
9	35	69	1.38	2.07	2.42	2.76	3.11	3.45	4.14	4.83	5.52	6.90
10	38	77	1.53	2.31	2.68	3.07	3.45	3.83	4.60	5.37	6.13	7.67
11	42	84	1.69	2.53	2.95	3.37	3.80	4.22	5.06	5.90	6.75	8.43
Days.												
1	0	0	01	01	01	01	01	01	02	02	02	03
2	0	01	01	02	02	02	02	03	03	04	04	05
3	0	01	02	02	03	03	03	04	05	05	06	08
4	01	01	02	03	04	04	05	05	06	07	08	10
5	01	01	03	04	04	05	06	06	08	09	10	13
6	01	02	03	05	05	06	07	08	09	11	12	15
7	01	02	04	05	06	07	08	09	11	13	14	18
8	01	02	04	06	07	08	09	10	12	14	16	20
9	01	02	05	07	08	09	10	12	14	16	18	23
10	01	03	05	08	09	10	12	13	15	18	20	26
11	01	03	06	08	10	11	13	14	17	20	22	28
12	02	03	06	09	11	12	14	15	18	21	24	31
13	02	03	07	10	12	13	15	17	20	23	26	33
14	02	04	07	11	13	14	16	18	21	25	28	36
15	02	04	08	12	13	15	17	19	23	27	31	38
16	02	04	08	12	14	16	18	20	25	29	33	41
17	02	04	09	13	15	17	20	22	26	30	35	43
18	02	05	09	14	16	18	21	23	28	32	37	46
19	02	05	10	15	17	19	22	24	29	34	39	49
20	03	05	10	15	18	20	23	26	31	36	41	51
21	03	05	11	16	19	21	24	27	32	38	43	54
22	03	06	11	17	20	22	25	28	34	39	45	56
23	03	06	12	18	21	24	26	29	35	41	47	59
24	03	06	12	18	21	25	28	31	37	43	49	61
25	03	06	13	19	22	26	29	32	38	45	51	64
26	03	07	13	20	23	27	30	33	40	47	53	66
27	03	07	14	21	24	28	31	35	41	48	55	69
28	04	07	14	21	25	29	32	36	43	50	57	72
29	04	07	15	22	26	30	33	37	44	52	59	74
30	04	08	15	23	27	31	35	38	46	54	61	77
33	04	08	17	25	30	34	38	42	51	59	67	84
63	08	16	32	48	56	64	72	81	97	1.13	1.29	1.61
93	12	24	48	71	83	95	1.07	1.19	1.43	1.66	1.90	2.38

93 Dollars.

Years.	½ per ct.	1 per ct.	2 per ct.	3 per ct.	3½ per ct.	4 per ct.	4½ per ct.	5 per ct.	6 per ct.	7 per ct.	8 per ct.	10 per ct.
1	47	93	1.86	2.79	3.26	3.72	4.19	4.65	5.58	6.51	7.44	9.30
2	93	1.86	3.72	5.58	6.51	7.44	8.37	9.30	11.16	13.02	14.88	18.60
3	1.40	2.79	5.58	8.37	9.77	11.16	12.56	13.95	16.74	19.53	22.32	27.90
4	1.86	3.72	7.44	11.16	13.02	14.88	16.74	18.60	22.32	26.04	29.76	37.20
5	2.33	4.65	9.30	13.95	16.28	18.60	20.93	23.25	27.90	32.55	37.20	46.50
Months.												
1	04	08	16	23	27	31	35	39	47	54	62	78
2	08	16	31	47	54	62	70	78	93	1.09	1.24	1.55
3	12	23	47	70	81	93	1.05	1.16	1.40	1.63	1.86	2.33
4	16	31	62	93	1.09	1.24	1.40	1.55	1.86	2.17	2.48	3.10
5	19	39	78	1.16	1.36	1.55	1.74	1.94	2.33	2.71	3.10	3.88
6	23	47	93	1.40	1.63	1.86	2.09	2.33	2.79	3.26	3.72	4.65
7	27	54	1.09	1.63	1.90	2.17	2.44	2.71	3.26	3.80	4.34	5.43
8	31	62	1.24	1.86	2.17	2.48	2.79	3.10	3.72	4.34	4.96	6.20
9	35	70	1.40	2.09	2.44	2.79	3.14	3.49	4.19	4.88	5.58	6.98
10	39	78	1.55	2.33	2.71	3.10	3.49	3.88	4.65	5.43	6.20	7.75
11	43	85	1.71	2.56	2.98	3.41	3.83	4.26	5.12	5.97	6.82	8.53
Days.												
1	0	0	01	01	01	01	01	01	02	02	02	03
2	0	01	01	02	02	02	02	03	03	04	04	05
3	0	01	02	02	03	03	03	04	05	05	06	08
4	01	01	02	03	04	04	05	05	06	07	08	10
5	01	01	03	04	05	05	06	06	08	09	10	13
6	01	02	03	05	05	06	07	08	09	11	12	16
7	01	02	04	05	06	07	08	09	11	13	14	18
8	01	02	04	06	07	08	09	10	12	14	17	21
9	01	02	05	07	08	09	10	12	14	16	19	23
10	01	03	05	08	09	10	12	13	16	18	21	26
11	01	03	06	09	10	11	13	14	17	20	23	28
12	02	03	06	09	11	12	14	16	19	22	25	31
13	02	03	07	10	12	13	15	17	20	24	27	34
14	02	04	07	11	13	14	16	18	22	25	29	36
15	02	04	08	12	14	16	17	19	23	27	31	39
16	02	04	08	12	14	17	19	21	25	29	33	41
17	02	04	09	13	15	18	20	22	26	31	35	44
18	02	05	09	14	16	19	21	23	28	33	37	47
19	02	05	10	15	17	20	22	25	29	34	39	49
20	03	05	10	16	18	21	23	26	31	36	41	52
21	03	05	11	16	19	22	24	27	33	38	43	54
22	03	06	11	17	20	23	26	28	34	40	45	57
23	03	06	12	18	21	24	27	30	36	42	48	59
24	03	06	12	19	22	25	28	31	37	43	50	62
25	03	06	13	19	23	26	29	32	39	45	52	65
26	03	07	13	20	24	27	30	34	40	47	54	67
27	03	07	14	21	24	28	31	35	42	49	56	70
28	04	07	14	22	25	29	33	36	43	51	58	72
29	04	07	15	22	26	30	34	37	45	52	60	75
30	04	08	16	23	27	31	35	39	47	54	62	78
33	04	09	17	26	30	34	38	43	51	60	68	85
63	08	16	33	49	57	65	73	81	98	1.14	1.30	1.63
93	12	24	48	72	84	96	1.08	1.20	1.44	1.68	1.92	2.40

94 Dollars.

	½ per ct.	1 per ct.	2 per ct.	3 per ct.	3½ per ct.	4 per ct.	4½ per ct.	5 per ct.	6 per ct.	7 per ct.	8 per ct.	10 per ct.
Years.												
1	47	94	1.88	2.82	3.29	3.76	4.23	4.70	5.64	6.58	7.52	9.40
2	94	1.88	3.76	5.64	6.58	7.52	8.46	9.40	11.28	13.16	15.04	18.80
3	1.41	2.82	5.64	8.46	9.87	11.28	12.69	14.10	16.92	19.74	22.56	28.20
4	1.88	3.76	7.52	11.28	13.16	15.04	16.92	18.80	22.56	26.32	30.08	37.60
5	2.35	4.70	9.40	14.10	16.45	18.80	21.15	23.50	28.20	32.90	37.60	47.00
Months.												
1	04	08	16	24	27	31	35	39	47	55	63	78
2	08	16	31	47	55	63	71	78	94	1.10	1.25	1.57
3	12	24	47	71	82	94	1.06	1.18	1.41	1.65	1.88	2.35
4	16	31	63	94	1.10	1.25	1.41	1.57	1.88	2.19	2.51	3.13
5	20	39	78	1.18	1.37	1.57	1.76	1.96	2.35	2.74	3.13	3.92
6	24	47	94	1.41	1.65	1.88	2.12	2.35	2.82	3.29	3.76	4.70
7	27	55	1.10	1.65	1.92	2.19	2.47	2.74	3.29	3.84	4.39	5.48
8	31	63	1.25	1.88	2.19	2.51	2.82	3.13	3.76	4.39	5.01	6.27
9	35	71	1.41	2.12	2.47	2.82·	3.17	3.53	4.23	4.94	5.64	7.05
10	39	78	1.57	2.35	2.74	3.13	3.53	3.92	4.70	5.48	6.27	7.83
11	43	86	1.72	2.59	3.02	3.45	3.88	4.31	5.17	6.03	6.89	8.62
Days.												
1	0	0	01	01	01	01	01	01	02	02	02	03
2	0	01	01	02	02	02	02	03	03	04	04	05
3	0	01	02	02	03	03	04	04	05	05	06	08
4	01	01	02	03	04	04	05	05	06	07	08	10
5	01	01	03	04	05	05	06	07	08	09	10	13
6	01	02	03	05	05	06	07	08	09	11	13	16
7	01	02	04	05	06	07	08	09	11	13	15	18
8	01	02	04	06	07	08	09	10	13	15	17	21
9	01	02	05	07	08	09	11	12	14	16	19	24
10	01	03	05	08	09	10	12	13	16	18	21	26
11	01	03	06	09	10	11	13	14	17	20	23	29
12	02	03	06	09	11	13	14	16	19	22	25	31
13	02	03	07	10	12	14	15	17	20	24	27	34
14	02	04	07	11	13	15	16	18	22	26	29	37
15	02	04	08	12	14	16	18	20	24	27	31	39
16	02	04	08	13	15	17	19	21	25	29	33	42
17	02	04	09	13	16	18	20	22	27	31	36	44
18	02	05	09	14	16	19	21	24	28	33	38	47
19	02	05	10	15	17	20	22	25	30	35	40	50
20	03	05	10	16	18	21	24	26	31	37	42	52
21	03	05	11	16	19	22	25	27	33	38	44	55
22	03	06	11	17	20	23	26	29	34	40	46	57
23	03	06	12	18	21	24	27	30	36	42	48	60
24	03	06	13	19	22	25	28	31	38	44	50	63
25	03	07	13	20	23	26	29	33	39	46	52	65
26	03	07	14	20	24	27	31	34	41	48	54	68
27	04	07	14	21	25	28	32	35	42	49	56	71
28	04	07	15	22	26	29	33	37	44	51	58	73
29	04	08	15	23	27	30	34	38	45	53	61	76
30	04	08	16	24	27	31	35	39	47	55	63	78
33	04	09	17	26	30	34	39	43	52	60	69	86
63	08	16	33	49	58	66	74	82	99	1.15	1.32	1.65
93	12	24	49	73	85	97	1.09	1.21	1.46	1.70	1.94	2.43

95 Dollars.

Years.	½ per ct.	1 per ct.	2 per ct.	3 per ct.	3½ per ct.	4 per ct.	4½ per ct.	5 per ct.	6 per ct.	7 per ct.	8 per ct.	10 per ct.
1	48	95	1.90	2.85	3.33	3.80	4.28	4.75	5.70	6.65	7.60	9.50
2	95	1.90	3.80	5.70	6.65	7.60	8.55	9.50	11.40	13.30	15.20	19.00
3	1.42	2.85	5.70	8.55	9.98	11.40	12.83	14.25	17.10	19.95	22.80	28.50
4	1.90	3.80	7.60	11.40	13.30	15.20	17.10	19.00	22.80	26.60	30.40	38.00
5	2.38	4.75	9.50	14.25	16.63	19.00	21.38	23.75	28.50	33.25	38.00	47.50
Months.												
1	04	08	16	24	28	32	36	40	48	55	63	79
2	08	16	32	48	55	63	71	79	95	1.11	1.27	1.58
3	12	24	48	71	83	95	1.07	1.19	1.43	1.66	1.90	2.38
4	16	32	63	95	1.11	1.27	1.43	1.58	1.90	2.22	2.53	3.17
5	20	40	79	1.19	1.39	1.58	1.78	1.98	2.38	2.77	3.17	3.96
6	24	48	95	1.43	1.66	1.90	2.14	2.38	2.85	3.33	3.80	4.76
7	28	55	1.11	1.66	1.94	2.22	2.49	2.77	3.33	3.88	4.43	5.54
8	32	63	1.27	1.90	2.22	2.53	2.85	3.17	3.80	4.43	5.07	6.33
9	36	71	1.43	2.14	2.49	2.85	3.21	3.56	4.28	4.99	5.70	7.13
10	40	79	1.58	2.38	2.77	3.17	3.56	3.96	4.75	5.54	6.33	7.92
11	44	87	1.74	2.61	3.05	3.48	3.92	4.35	5.23	6.10	6.97	8.71
Days.												
1	0	0	01	01	01	01	01	01	02	02	02	03
2	0	01	01	02	02	02	02	03	03	04	04	05
3	0	01	02	03	03	03	04	04	05	06	06	08
4	01	01	02	03	04	04	05	05	06	07	08	11
5	01	01	03	04	05	05	06	07	08	09	11	13
6	01	02	03	05	06	06	07	08	10	11	13	16
7	01	02	04	06	06	07	08	09	11	13	15	18
8	01	02	04	06	07	08	10	11	13	15	17	21
9	01	02	05	07	08	10	11	12	14	17	19	24
10	01	03	05	08	09	11	12	13	16	18	21	26
11	01	03	06	09	10	12	13	15	17	20	23	29
12	02	03	06	10	11	13	14	16	19	22	25	32
13	02	03	07	10	12	14	15	17	21	24	27	34
14	02	04	07	11	13	15	17	18	22	26	30	37
15	02	04	08	12	14	16	18	20	24	28	32	40
16	02	04	08	13	15	17	19	21	25	30	34	42
17	02	04	09	13	16	18	20	22	27	31	36	45
18	02	05	10	14	17	19	21	24	29	33	38	48
19	03	05	10	15	18	20	23	25	30	35	40	50
20	03	05	11	16	18	21	24	26	32	37	42	53
21	03	06	11	17	19	22	25	28	33	39	44	55
22	03	06	12	17	20	23	26	29	35	41	46	58
23	03	06	12	18	21	24	27	30	36	42	49	61
24	03	06	13	19	22	25	29	32	38	44	51	63
25	03	07	13	20	23	26	30	33	40	46	53	66
26	03	07	14	21	24	27	31	34	41	48	55	69
27	04	07	14	21	25	29	32	36	43	50	57	71
28	04	07	15	22	26	30	33	37	44	52	59	74
29	04	08	15	23	27	31	34	38	46	54	61	77
30	04	08	16	24	28	32	36	40	48	55	63	79
33	04	09	17	26	30	35	39	44	52	61	70	87
63	08	17	33	50	58	67	75	83	1.00	1.16	1.33	1.66
93	12	25	49	74	86	98	1.10	1.23	1.47	1.72	1.96	2.45

96 Dollars.

Years.	½ per ct.	1 per ct.	2 per ct.	3 per ct.	3½ per ct.	4 per ct.	4½ per ct.	5 per ct.	6 per ct.	7 per ct.	8 per ct.	10 per ct.
1	48	96	1.92	2.88	3.36	3.84	4.32	4.80	5.76	6.72	7.68	9.60
2	96	1.92	3.84	5.76	6.72	7.68	8.64	9.60	11.52	13.44	15.36	19.20
3	1.44	2.88	5.76	8.64	10.08	11.52	12.96	14.40	17.28	20.16	23.04	28.80
4	1.92	3.84	7.68	11.52	13.44	15.36	17.28	19.20	23.04	26.88	30.72	38.40
5	2.40	4.80	9.60	14.40	16.80	19.20	21.60	24.00	28.80	33.60	38.40	48.00
Months.												
1	04	08	16	24	28	32	36	40	48	56	64	80
2	08	16	32	48	56	64	72	80	96	1.12	1.28	1.60
3	12	24	48	72	84	96	1.08	1.20	1.44	1.68	1.92	2.40
4	16	32	64	96	1.12	1.28	1.44	1.60	1.92	2.24	2.56	3.20
5	20	40	80	1.20	1.40	1.60	1.80	2.00	2.40	2.80	3.20	4.00
6	24	48	96	1.44	1.68	1.92	2.16	2.40	2.88	3.36	3.84	4.80
7	28	56	1.12	1.68	1.96	2.24	2.52	2.80	3.36	3.92	4.44	5.60
8	32	64	1.28	1.92	2.24	2.56	2.88	3.20	3.84	4.48	5.12	6.40
9	36	72	1.44	2.16	2.52	2.88	3.24	3.60	4.32	5.04	5.76	7.20
10	40	80	1.60	2.40	2.80	3.20	3.60	4.00	4.80	5.60	6.40	8.00
11	44	88	1.76	2.64	3.08	3.52	3.96	4.40	5.28	6.16	7.04	8.80
Days.												
1	0	0	01	01	01	01	01	01	02	02	02	03
2	0	01	01	02	02	02	02	03	03	04	04	05
3	0	01	02	02	03	03	04	04	05	06	06	08
4	01	01	02	03	04	04	05	05	06	07	09	11
5	01	01	03	04	05	05	06	07	08	09	11	13
6	01	02	03	05	06	06	07	08	10	11	13	16
7	01	02	04	06	07	07	08	09	11	13	15	19
8	01	02	04	06	07	09	10	11	13	15	17	21
9	01	02	05	07	08	10	11	12	14	17	19	24
10	01	03	05	08	09	11	12	13	16	19	21	27
11	01	03	06	09	10	12	13	15	18	21	23	29
12	02	03	06	10	11	13	14	16	19	22	26	32
13	02	03	07	10	12	14	16	17	21	24	28	35
14	02	04	07	11	13	15	17	19	22	26	30	37
15	02	04	08	12	14	16	18	20	24	28	32	40
16	02	04	09	13	15	17	19	21	26	30	34	43
17	02	05	09	14	16	18	20	23	27	32	36	45
18	02	05	10	14	17	19	22	24	29	34	38	48
19	03	05	10	15	18	20	23	25	30	35	41	51
20	03	05	11	16	19	21	24	27	32	37	43	53
21	03	06	11	17	20	22	25	28	34	39	45	56
22	03	06	12	18	21	23	26	29	35	41	47	59
23	03	06	12	18	21	25	28	31	37	43	49	61
24	03	06	13	19	22	26	29	32	38	45	51	64
25	03	07	13	20	23	27	30	33	40	47	53	67
26	03	07	14	21	24	28	31	35	42	49	55	69
27	04	07	14	22	25	29	32	36	43	50	58	72
28	04	07	15	22	26	30	34	37	45	52	60	75
29	04	08	15	23	27	31	35	39	46	54	62	77
30	04	08	16	24	28	32	36	40	48	56	64	80
33	04	09	18	26	31	35	40	44	53	62	70	88
63	08	17	34	50	59	67	76	84	1.01	1.18	1.34	1.68
93	12	25	50	74	87	99	1.12	1.24	1.49	1.74	1.98	2.48

97 Dollars.

Years.	½ per ct.	1 per ct.	2 per ct.	3 per ct.	3½ per ct.	4 per ct.	4½ per ct.	5 per ct.	6 per ct.	7 per ct.	8 per ct.	10 per ct.
1	49	97	1.94	2.91	3.40	3.88	4.37	4.85	5.82	6.79	7.76	9.70
2	97	1.94	3.88	5.82	6.79	7.76	8.73	9.70	11.64	13.58	15.52	19.40
3	1.46	2.91	5.82	8.73	10.19	11.64	13.10	14.55	17.46	20.37	23.28	29.10
4	1.94	3.88	7.76	11.64	13.58	15.52	17.46	19.40	23.28	27.16	31.04	38.80
5	2.43	4.85	9.70	14.55	16.98	19.40	21.83	24.25	29.10	33.95	38.80	48.50
Months.												
1	04	08	16	24	28	32	36	40	49	57	65	81
2	08	16	32	49	57	65	73	81	97	1.13	1.29	1.62
3	12	24	49	73	85	97	1.09	1.21	1.46	1.70	1.94	2.43
4	16	32	65	97	1.13	1.29	1.46	1.62	1.94	2.26	2.59	3.23
5	20	40	81	1.21	1.41	1.62	1.82	2.02	2.43	2.83	3.23	4.04
6	24	49	97	1.46	1.70	1.94	2.18	2.43	2.91	3.40	3.88	4.85
7	28	57	1.13	1.70	1.98	2.26	2.55	2.83	3.40	3.96	4.53	5.66
8	32	65	1.29	1.94	2.26	2.59	2.91	3.23	3.88	4.53	5.17	6.47
9	36	73	1.46	2.18	2.55	2.91	3.27	3.64	4.37	5.09	5.82	7.28
10	40	81	1.62	2.43	2.83	3.23	3.64	4.04	4.85	5.66	6.47	8.08
11	44	89	1.78	2.67	3.11	3.56	4.00	4.45	5.34	6.22	7.11	8.89
Days.												
1	0	0	01	01	01	01	01	01	02	02	02	03
2	0	01	01	02	02	02	02	03	03	04	04	05
3	0	01	02	02	03	03	04	04	05	06	06	08
4	01	01	02	03	04	04	05	05	06	08	09	11
5	01	01	03	04	05	05	06	07	08	09	11	13
6	01	02	03	05	06	06	07	08	10	11	13	16
7	01	02	04	06	07	08	08	09	11	13	15	19
8	01	02	04	06	08	09	10	11	13	15	17	22
9	01	02	05	07	08	10	11	12	15	17	19	24
10	01	03	05	08	09	11	12	13	16	19	22	27
11	01	03	06	09	10	12	13	15	18	21	24	30
12	02	03	06	10	11	13	15	16	19	23	26	32
13	02	04	07	11	12	14	16	18	21	25	28	35
14	02	04	08	11	13	15	17	19	23	26	30	38
15	02	04	08	12	14	16	18	20	24	28	32	40
16	02	04	09	13	15	17	19	22	26	30	34	43
17	02	05	09	14	16	18	21	23	27	32	37	46
18	02	05	10	15	17	19	22	24	29	34	39	49
19	03	05	10	15	18	20	23	26	31	36	41	51
20	03	05	11	16	19	22	24	27	32	38	43	54
21	03	06	11	17	20	23	25	28	34	40	45	57
22	03	06	12	18	21	24	27	30	36	41	47	59
23	03	06	12	19	22	25	28	31	37	43	50	62
24	03	06	13	19	23	26	29	32	39	45	52	65
25	03	07	13	20	24	27	30	34	40	47	54	67
26	04	07	14	21	25	28	32	35	42	49	56	70
27	04	07	15	22	25	29	33	36	44	51	58	73
28	04	08	15	23	26	30	34	38	45	53	60	75
29	04	08	16	23	27	31	35	39	47	55	63	78
30	04	08	16	24	28	32	36	40	49	57	65	81
33	04	09	18	27	31	36	40	44	54	62	71	89
63	08	17	34	51	59	68	76	85	1.02	1.19	1.35	1.70
93	13	25	50	75	88	1.00	1.13	1.25	1.50	1.75	2.00	2.51

98 Dollars.

Years.	½ per ct.	1 per ct.	2 per ct.	3 per ct.	3½ per ct.	4 per ct.	4½ per ct.	5 per ct.	6 per ct.	7 per ct.	8 per ct.	10 per ct.
1	49	98	1.96	2.94	3.43	3.92	4.41	4.90	5.88	6.86	7.84	9.80
2	98	1.96	3.92	5.88	6.86	7.84	8.82	9.80	11.76	13.72	15.68	19.60
3	1.47	2.94	5.88	8.82	10.29	11.76	13.23	14.70	17.64	20.58	23.52	29.40
4	1.96	3.92	7.84	11.76	13.72	15.68	17.64	19.60	23.52	27.44	31.36	39.20
5	2.45	4.90	9.80	14.70	17.15	19.60	22.05	24.50	29.40	34.30	39.20	49.00
Months.												
1	04	08	16	25	29	33	37	41	49	57	65	82
2	08	16	33	49	57	65	74	82	98	1.14	1.31	1.63
3	12	25	49	74	86	98	1.10	1.23	1.47	1.72	1.96	2.45
4	16	33	65	98	1.14	1.31	1.47	1.63	1.96	2.29	2.61	3.27
5	20	41	82	1.23	1.43	1.63	1.84	2.04	2.45	2.86	3.27	4.08
6	25	49	98	1.47	1.72	1.96	2.21	2.45	2.94	3.43	3.92	4.90
7	29	57	1.14	1.72	2.00	2.29	2.57	2.86	3.43	4.00	4.57	5.72
8	33	65	1.31	1.96	2.29	2.61	2.94	3.27	3.93	4.57	5.23	6.53
9	37	74	1.47	2.21	2.57	2.94	3.31	3.68	4.41	5.15	5.88	7.35
10	41	82	1.63	2.45	2.86	3.27	3.68	4.08	4.90	5.72	6.53	8.17
11	45	90	1.80	2.70	3.14	3.59	4.04	4.49	5.39	6.29	7.19	8.98
Days.												
1	0	0	01	01	01	01	01	01	02	02	02	03
2	0	01	01	02	02	02	02	03	03	04	04	05
3	0	01	02	02	03	03	04	04	05	06	07	08
4	01	01	02	03	04	04	05	05	07	08	09	11
5	01	01	03	04	05	05	06	07	08	10	11	14
6	01	02	03	05	06	07	07	08	10	11	13	16
7	01	02	04	06	07	08	09	10	11	13	15	19
8	01	02	04	07	08	09	10	11	13	15	17	22
9	01	02	05	07	09	10	11	12	15	17	20	25
10	01	03	05	08	10	11	12	14	16	19	22	27
11	01	03	06	09	10	12	13	15	18	21	24	30
12	02	03	07	10	11	13	15	16	20	23	26	33
13	02	04	07	11	12	14	16	18	21	25	28	35
14	02	04	08	11	13	15	17	19	23	27	30	38
15	02	04	08	12	14	16	18	20	25	29	33	41
16	02	04	09	13	15	17	20	22	26	30	35	44
17	02	05	09	14	16	19	21	23	28	32	37	46
18	02	05	10	15	17	20	22	25	29	34	39	49
19	03	05	10	16	18	21	23	26	31	36	41	52
20	03	05	11	16	19	22	25	27	33	38	44	54
21	03	06	11	17	20	23	26	29	34	40	46	57
22	03	06	12	18	21	24	27	30	36	42	48	60
23	03	06	13	19	22	25	28	31	38	44	50	63
24	03	07	13	20	23	26	29	33	39	46	52	65
25	03	07	14	20	24	27	31	34	41	48	54	68
26	04	07	14	21	25	28	32	35	42	50	57	71
27	04	07	15	22	26	29	33	36	44	51	59	74
28	04	08	15	23	27	30	34	38	46	53	61	76
29	04	08	16	24	28	32	36	39	47	55	63	79
30	04	08	16	25	29	33	37	41	49	57	65	82
33	04	09	18	27	31	36	40	45	54	63	72	90
63	09	17	34	51	60	69	77	86	1.03	1.20	1.37	1.72
93	13	25	51	76	89	1.01	1.14	1.27	1.52	1.77	2.03	2.53

99 Dollars.

	½ per ct.	1 per ct.	2 per ct.	3 per ct.	3½ per ct.	4 per ct.	4½ per ct.	5 per ct.	6 per ct.	7 per ct.	8 per ct.	10 per ct.
Years.												
1	50	99	1.98	2.97	3.47	3.96	4.46	4.95	5.94	6.93	7.92	9.90
2	99	1.98	3.96	5.94	6.93	7.92	8.91	9.90	11.88	13.86	15.84	19.80
3	1.49	2.97	5.94	8.91	10.40	11.88	13.37	14.85	17.82	20.79	23.76	29.70
4	1.98	3.96	7.92	11.88	13.86	15.84	17.82	19.80	23.76	27.72	31.68	39.60
5	2.48	4.95	9.90	14.85	17.33	19.80	22.28	24.75	29.70	34.65	39.60	49.50
Months.												
1	04	08	17	25	29	33	37	41	50	58	66	83
2	08	17	33	50	58	66	74	83	99	1.16	1.32	1.65
3	12	25	50	74	87	99	1.11	1.24	1.49	1.73	1.98	2.48
4	17	33	66	99	1.16	1.32	1.49	1.65	1.98	2.31	2.64	3.30
5	21	41	83	1.24	1.44	1.65	1.86	2.06	2.48	2.89	3.30	4.13
6	25	50	99	1.49	1.73	1.98	2.23	2.48	2.97	3.47	3.96	4.95
7	29	58	1.16	1.73	2.02	2.31	2.60	2.89	3.47	4.09	4.62	5.78
8	33	66	1.32	1.98	2.31	2.64	2.97	3.30	3.96	4.62	5.28	6.60
9	37	74	1.49	2.23	2.60	2.97	3.34	3.71	4.46	5.20	5.94	7.43
10	41	83	1.65	2.48	2.89	3.30	3.71	4.13	4.95	5.78	6.60	8.25
11	45	91	1.82	2.72	3.18	3.63	4.08	4.54	5.45	6.35	7.26	9.08
Days.												
1	0	0	01	01	01	01	01	01	02	02	02	03
2	0	01	01	02	02	02	02	03	03	04	04	06
3	0	01	02	02	03	03	04	04	05	06	07	08
4	01	01	02	03	04	04	05	06	07	08	09	11
5	01	01	03	04	05	06	06	07	08	10	11	14
6	01	02	03	05	06	07	07	08	10	12	13	17
7	01	02	04	06	07	08	09	10	12	13	15	19
8	01	02	04	07	08	09	10	11	13	15	18	22
9	01	02	05	07	09	10	11	12	15	17	20	25
10	01	03	06	08	10	11	12	14	17	19	22	28
11	02	03	06	09	11	12	14	15	18	21	24	30
12	02	03	07	10	12	13	15	17	20	23	26	33
13	02	04	07	11	13	14	16	18	21	25	29	36
14	02	04	08	12	13	15	17	19	23	27	31	39
15	02	04	08	12	14	17	19	21	25	29	33	41
16	02	04	09	13	15	18	20	22	26	31	35	44
17	02	05	09	14	16	19	21	23	28	33	37	47
18	02	05	10	15	17	20	22	25	30	35	40	50
19	03	05	10	16	18	21	24	26	31	37	42	52
20	03	06	11	17	19	22	25	28	33	39	44	55
21	03	06	12	17	20	23	26	29	35	40	46	58
22	03	06	12	18	21	24	27	30	36	42	48	61
23	03	06	13	19	22	25	28	32	38	44	51	63
24	03	07	13	20	23	26	30	33	40	46	53	66
25	03	07	14	21	24	28	31	34	41	48	55	69
26	04	07	14	21	25	29	32	36	43	50	57	72
27	04	07	15	22	26	30	33	37	45	52	59	74
28	04	08	15	23	27	31	35	39	46	54	62	77
29	04	08	16	24	28	32	36	40	48	56	64	80
30	04	08	17	25	29	33	37	41	50	58	66	83
33	05	09	18	27	32	36	41	45	54	64	73	91
63	09	17	35	52	61	69	78	87	1.04	1.21	1.39	1.73
93	13	26	51	77	90	1.02	1.15	1.28	1.53	1.79	2.05	2.56

100 Dollars.

Years.	½ per ct.	1 per ct.	2 per ct.	3 per ct.	3½ per ct.	4 per ct.	4½ per ct.	5 per ct.	6 per ct.	7 per ct.	8 per ct.	10 per ct.
1	50	1.00	2.00	3.00	3.50	4.00	4.50	5.00	6.00	7.00	8.00	10.00
2	1.00	2.00	4.00	6.00	7.00	8.00	9.00	10.00	12.00	14.00	16.00	20.00
3	1.50	3.00	6.00	9.00	10.50	12.00	13.50	15.00	18.00	21.00	24.00	30 00
4	2.00	4.00	8.00	12.00	14.00	16.00	18.00	20.00	24.00	28.00	32.00	40.00
5	2.50	5.00	10.00	15.00	17.50	20.00	22.50	25.00	30.00	35.00	40.00	50.00
Months.												
1	04	08	17	25	29	33	38	42	50	58	67	83
2	08	17	33	50	58	67	75	83	1.00	1.17	1.33	1.67
3	13	25	50	75	88	1.00	1.13	1.25	1.50	1.75	2.00	2.50
4	17	33	67	1.00	1.17	1.33	1.50	1.67	2.00	2.33	2.67	3.33
5	21	42	83	1.25	1.46	1.67	1.88	2.08	2.50	2.92	3.33	4.17
6	25	50	1.00	1.50	1.75	2.00	2.25	2.50	3.00	3.50	4.00	5.00
7	29	58	1.17	1.75	2.04	2.33	2.62	2.92	3.50	4.08	4.67	5.83
8	33	67	1.33	2.00	2.33	2.67	3.00	3.33	4.00	4.67	5.33	6.67
9	38	75	1.50	2.25	2.63	3.00	3.38	3.75	4.50	5.25	6.00	7.50
10	42	83	1.67	2.50	2.92	3.33	3.75	4.17	5.00	5.83	6.67	8.33
11	46	92	1.83	2.75	3.21	3.67	4.13	4.58	5.50	6.42	7.33	9.17
Days.												
1	0	0	01	01	01	01	01	01	02	02	02	03
2	0	01	01	02	02	02	03	03	03	04	04	06
3	0	01	02	03	03	03	04	04	05	06	07	08
4	01	01	02	03	04	04	05	06	07	08	09	11
5	01	01	03	04	05	06	06	07	08	10	11	14
6	01	02	03	05	06	07	08	08	10	12	13	17
7	01	02	04	06	07	08	09	10	12	14	16	19
8	01	02	04	07	08	09	10	11	13	16	18	22
9	01	03	05	08	09	10	11	13	15	18	20	25
10	01	03	06	08	10	11	13	14	17	19	22	28
11	02	03	06	09	11	12	14	15	18	21	24	31
12	02	03	07	10	12	13	15	17	20	23	27	33
13	02	04	07	11	13	14	16	18	22	25	29	36
14	02	04	08	12	14	16	18	19	23	27	31	39
15	02	04	08	13	15	17	19	21	25	29	33	42
16	02	04	09	13	16	18	20	22	27	31	36	44
17	02	05	09	14	17	19	21	24	28	33	38	47
18	03	05	10	15	18	20	23	25	30	35	40	50
19	03	05	11	16	18	21	24	26	32	37	42	53
20	03	06	11	17	19	22	25	28	33	39	44	56
21	03	06	12	18	20	23	26	29	35	41	47	58
22	03	06	12	18	21	24	28	31	37	43	49	61
23	03	06	13	19	22	26	29	32	38	45	51	64
24	03	07	13	20	23	27	30	33	40	47	53	67
25	03	07	14	21	24	28	31	35	42	49	56	69
26	04	07	14	22	25	29	33	36	43	51	58	72
27	04	08	15	23	26	30	34	38	45	53	60	75
28	04	08	16	23	27	31	35	39	47	54	62	78
29	04	08	16	24	28	32	36	40	48	56	64	81
30	04	08	17	25	29	33	38	42	50	58	67	83
33	05	09	18	28	32	37	42	46	55	64	73	92
63	09	18	35	53	61	70	79	88	1.05	1.23	1.40	1.75
93	13	26	52	78	90	1.03	1.17	1.29	1.55	1.81	2.07	2.58

200 Dollars.

	½ per ct.	1 per ct.	2 per ct.	3 per ct.	3½ per ct.	4 per ct.	4½ per ct.	5 per ct.	6 per ct.	7 per ct.	8 per ct.	10 per ct.
Years.												
1	1.00	2.00	4.00	6.00	7.00	8.00	9.00	10.00	12.00	14.00	16.00	20.00
2	2.00	4.00	8.00	12.00	14.00	16.00	18.00	20.00	24.00	28.00	32.00	40.00
3	3.00	6.00	12.00	18.00	21.00	24.00	27.00	30.00	36.00	42.00	48.00	60.00
4	4.00	8.00	16.00	24.00	28.00	32.00	36.00	40.00	48.00	56.00	64.00	80.00
5	5.00	10.00	20.00	30.00	35.00	40.00	45.00	50.00	60.00	70.00	80.00	100.00
Months.												
1	08	17	33	50	58	67	75	83	1.00	1.17	1.33	1.67
2	17	33	67	1.00	1.17	1.33	1.50	1.67	2.00	2.33	2.67	3.33
3	25	50	1.00	1.50	1.75	2.00	2.25	2.50	3.00	3.50	4.00	5.00
4	33	67	1.33	2.00	2.33	2.67	3.00	3.33	4.00	4.67	5.33	6.67
5	42	83	1.67	2.50	2.92	3.33	3.75	4.17	5.00	5.83	6.67	8.33
6	50	1.00	2.00	3.00	3.50	4.00	4.50	5.00	6.00	7.00	8.00	10.00
7	58	1.17	2.33	3.50	4.08	4.67	5.25	5.83	7.00	8.17	9.33	11.67
8	67	1.33	2.67	4.00	4.67	5.33	6.00	6.67	8.00	9.33	10.67	13.33
9	75	1.50	3.00	4.50	5.25	6.00	6.75	7.50	9.00	10.50	12.00	15.00
10	83	1.67	3.33	5.00	5.83	6.67	7.50	8.33	10.00	11.67	13.33	16.67
11	92	1.83	3.67	5.50	6.42	7.33	8.25	9.17	11.00	12.83	14.67	18.33
Days.												
1	0	01	01	02	02	02	03	03	03	04	04	06
2	01	01	02	03	04	04	05	06	07	08	09	11
3	01	02	03	05	06	07	08	08	10	12	13	17
4	01	02	04	07	08	09	10	11	13	16	18	22
5	01	03	06	08	10	11	13	14	17	19	22	28
6	02	03	07	10	12	13	15	17	20	23	27	33
7	02	04	08	12	14	16	18	19	23	27	31	39
8	02	04	09	13	16	18	20	22	27	31	36	44
9	03	05	10	15	18	20	23	25	30	35	40	50
10	03	06	11	17	19	22	25	28	33	39	44	56
11	03	06	12	18	21	24	28	31	37	43	49	61
12	03	07	13	20	23	27	30	33	40	47	53	67
13	04	07	14	22	25	29	33	36	43	51	58	72
14	04	08	16	23	27	31	35	39	47	54	62	78
15	04	08	17	25	29	33	38	42	50	58	67	83
16	04	09	18	27	31	36	40	44	53	62	71	89
17	05	09	19	28	33	38	43	47	57	66	76	94
18	05	10	20	30	35	40	45	50	60	70	80	1.00
19	05	11	21	32	37	42	48	53	63	74	84	1.06
20	06	11	22	33	39	44	50	56	67	78	89	1.11
21	06	12	23	35	41	47	53	58	70	82	93	1.17
22	06	12	24	37	43	49	55	61	73	86	98	1.22
23	06	13	26	38	45	51	58	64	77	89	1.02	1.28
24	07	13	27	40	47	53	60	67	80	93	1.07	1.33
25	07	14	28	42	49	56	63	69	83	97	1.11	1.39
26	07	14	29	43	51	58	65	72	87	1.01	1.16	1.44
27	08	15	30	45	53	60	68	75	90	1.05	1.20	1.50
28	08	16	31	47	54	62	70	78	93	1.09	1.24	1.56
29	08	16	32	48	56	64	73	81	97	1.13	1.29	1.61
30	08	17	33	50	58	67	75	83	1.00	1.17	1.33	1.67
33	09	19	37	55	64	73	83	92	1.10	1.28	1.47	1.83
63	18	35	70	1.05	1.23	1.40	1.58	1.75	2.10	2.45	2.80	3.50
93	26	52	1.03	1.55	1.81	2.07	2.33	2.58	3.10	3.62	4.13	5.17

300 Dollars.

	½ per ct.	1 per ct.	2 per ct.	3 per ct.	3½ per ct.	4 per ct.	4½ per ct.	5 per ct.	6 per ct.	7 per ct.	8 per ct.	10 per ct.
Years.												
1	1.50	3.00	6.00	9.00	10.50	12.00	13.50	15.00	18.00	21.00	24.00	30.00
2	3.00	6.00	12.00	18.00	21.00	24.00	27.00	30.00	36.00	42.00	48.00	60.00
3	4.50	9.00	18.00	27.00	31.50	36.00	40.50	45.00	54.00	63.00	72.00	90.00
4	6.00	12.00	24.00	36.00	42.00	48.00	54.00	60.00	72.00	84.00	96.00	120.00
5	7.50	15.00	30.00	45.00	52.50	60.00	67.50	75.00	90.00	105.00	120.00	150.00
Months.												
1	13	25	50	75	88	1.00	1.13	1.25	1.50	1.75	2.00	2.50
2	25	50	1.00	1.50	1.75	2.00	2.25	2.50	3.00	3.50	4.00	5.00
3	38	75	1.50	2.25	2.63	3.00	3.38	3.75	4.50	5.25	6.00	7.50
4	50	1.00	2.00	3.00	3.50	4.00	4.50	5.00	6.00	7.00	8.00	10.00
5	63	1.25	2.50	3.75	4.38	5.00	5.63	6.25	7.50	8.75	10.00	12.50
6	75	1.50	3.00	4.50	5.25	6.00	6.75	7.50	9.00	10.50	12.00	15.00
7	88	1.75	3.50	5.25	6.13	7.00	7.88	8.75	10.50	12.25	14.00	17.50
8	1.00	2.00	4.00	6.00	7.00	8.00	9.00	10.00	12.00	14.00	16.00	20.00
9	1.13	2.25	4.50	6.75	7.88	9.00	10.13	11.25	13.50	15.75	18.00	22.50
10	1.25	2.50	5.00	7.50	8.75	10.00	11.25	12.50	15.00	17.50	20.00	25.00
11	1.38	2.75	5.50	8.25	9.62	11.00	12.38	13.75	16.50	19.25	22.00	27.50
Days.												
1	0	01	02	03	03	03	04	04	05	06	07	08
2	01	02	03	05	06	07	08	08	10	12	13	17
3	01	03	05	08	09	10	11	13	15	18	20	25
4	02	03	07	10	12	13	15	17	20	23	27	33
5	02	04	08	13	15	17	19	21	25	29	33	42
6	03	05	10	15	18	20	23	25	30	35	40	50
7	03	06	12	18	20	23	26	29	35	41	47	58
8	03	07	13	20	23	27	30	33	40	47	53	67
9	04	08	15	23	26	30	34	38	45	53	60	75
10	04	08	17	25	29	33	38	42	50	58	67	83
11	05	09	18	28	32	37	41	46	55	64	73	92
12	05	10	20	30	35	40	45	50	60	70	80	1.00
13	05	11	22	33	38	43	49	54	65	76	87	1.08
14	06	12	23	35	41	47	53	58	70	82	93	1.17
15	06	13	25	38	44	50	56	63	75	88	1.00	1.25
16	07	13	27	40	47	53	60	67	80	93	1.07	1.33
17	07	14	28	43	50	57	64	71	85	99	1.13	1.42
18	08	15	30	45	53	60	68	75	90	1.05	1.20	1.50
19	08	16	32	48	55	63	71	79	95	1.11	1.27	1.58
20	08	17	33	50	58	67	75	83	1.00	1.17	1.33	1.67
21	09	18	35	53	61	70	79	88	1.05	1.23	1.40	1.75
22	09	18	37	55	64	73	83	92	1.10	1.28	1.47	1.83
23	10	19	38	58	67	77	86	96	1.15	1.34	1.53	1.92
24	10	20	40	60	70	80	90	1.00	1.20	1.40	1.60	2.00
25	10	21	42	63	73	83	94	1.04	1.25	1.46	1.67	2.08
26	11	22	43	65	76	87	98	1.08	1.30	1.52	1.73	2.17
27	11	23	45	68	79	90	1.01	1.13	1.35	1.58	1.80	2.25
28	12	23	47	70	82	93	1.05	1.17	1.40	1.63	1.87	2.33
29	12	24	48	73	85	97	1.09	1.21	1.45	1.69	1.93	2.42
30	13	25	50	75	88	1.00	1.13	1.25	1.50	1.75	2.00	2.50
33	14	28	55	83	96	1.10	1.24	1.38	1.65	1.93	2.20	2.75
63	26	53	1.05	1.58	1.84	2.10	2.36	2.63	3.15	3.68	4.20	5.25
93	39	78	1.55	2.33	2.71	3.10	3.49	3.88	4.65	5.43	6.20	7.75

400 Dollars.

Years.	½ per ct.	1 per ct.	2 per ct.	3 per ct.	3½ per ct.	4 per ct.	4½ per ct.	5 per ct.	6 per ct.	7 per ct.	8 per ct.	10 per ct.
1	2.00	4.00	8.00	12.00	14.00	16.00	18.00	20.00	24.00	28.00	32.00	40.00
2	4.00	8.00	16.00	24.00	28.00	32.00	36.00	40.00	48.00	56.00	64.00	80.00
3	6.00	12.00	24.00	36.00	42.00	48.00	54.00	60.00	72.00	84.00	96.00	120.00
4	8.00	16.00	32.00	48.00	56.00	64.00	72.00	80.00	96.00	112.00	128.00	160.00
5	10.00	20.00	40.00	60.00	70.00	80.00	90.00	100.00	120.00	140.00	160.00	200.00

Months.												
1	17	33	67	1.00	1.17	1.33	1.50	1.67	2.00	2.33	2.67	3.33
2	33	67	1.33	2.00	2.33	2.67	3.00	3.33	4.00	4.67	5.33	6.67
3	50	1.00	2.00	3.00	3.50	4.00	4.50	5.00	6.00	7.00	8.00	10.00
4	67	1.33	2.67	4.00	4.67	5.33	6.00	6.67	8.00	9.33	10.67	13.33
5	83	1.67	3.33	5.00	5.83	6.67	7.50	8.33	10.00	11.67	13.33	16.67
6	1.00	2.00	4.00	6.00	7.00	8.00	9.00	10.00	12.00	14.00	16.00	20.00
7	1.17	2.33	4.67	7.00	8.17	9.33	10.50	11.67	14.00	16.33	18.67	23.33
8	1.33	2.67	5.33	8.00	9.33	10.67	12.00	13.33	16.00	18.67	21.33	26.67
9	1.50	3.00	6.00	9.00	10.50	12.00	13.50	15.00	18.00	21.00	24.00	30.00
10	1.67	3.33	6.67	10.00	11.67	13.33	15.00	16.67	20.00	23.33	26.67	33.33
11	1.83	3.67	7.33	11.00	12.83	14.67	16.50	18.33	22.00	25.67	29.33	36.67

Days.												
1	01	01	02	03	04	04	05	06	07	08	09	11
2	01	02	04	07	08	09	10	11	13	16	18	22
3	02	03	07	10	12	13	15	17	20	23	27	33
4	02	04	09	13	16	18	20	22	27	31	36	44
5	03	06	11	17	19	22	25	28	33	39	44	56
6	03	07	13	20	23	27	30	33	40	47	53	67
7	04	08	16	23	27	31	35	39	47	54	62	78
8	04	09	18	27	31	36	40	44	53	62	71	89
9	05	10	20	30	35	40	45	50	60	70	80	1.00
10	06	11	22	33	39	44	50	56	67	78	89	1.11
11	06	12	24	37	43	49	55	61	73	86	98	1.22
12	07	13	27	40	47	53	60	67	80	93	1.07	1.33
13	07	14	29	43	51	58	65	72	87	1.01	1.16	1.44
14	08	16	31	47	54	62	70	78	93	1.09	1.24	1.56
15	08	17	33	50	58	67	75	83	1.00	1.17	1.33	1.67
16	09	18	36	53	62	71	80	89	1.07	1.24	1.42	1.78
17	09	19	38	57	66	76	85	94	1.13	1.32	1.51	1.89
18	10	20	40	60	70	80	90	1.00	1.20	1.40	1.60	2.00
19	11	21	42	63	74	84	95	1.06	1.27	1.48	1.69	2.11
20	11	22	44	67	78	89	1.00	1.11	1.33	1.56	1.78	2.22
21	12	23	47	70	82	93	1.05	1.17	1.40	1.63	1.87	2.33
22	12	24	49	73	86	98	1.10	1.22	1.47	1.71	1.96	2.44
23	13	26	51	77	89	1.02	1.15	1.28	1.53	1.79	2.04	2.56
24	13	27	53	80	93	1.07	1.20	1.33	1.60	1.87	2.13	2.67
25	14	28	56	83	97	1.11	1.25	1.39	1.67	1.94	2.22	2.78
26	14	29	58	87	1.01	1.16	1.30	1.44	1.73	2.02	2.31	2.89
27	15	30	60	90	1.05	1.20	1.35	1.50	1.80	2.10	2.40	3.00
28	16	31	62	93	1.09	1.24	1.40	1.56	1.87	2.18	2.49	3.11
29	16	32	64	97	1.13	1.29	1.45	1.61	1.93	2.26	2.58	3.22
30	17	33	67	1.00	1.17	1.33	1.50	1.67	2.00	2.33	2.67	3.33
33	18	37	73	1.10	1.28	1.47	1.65	1.83	2.20	2.57	2.93	3.67
63	35	70	1.40	2.10	2.45	2.80	3.15	3.50	4.20	4.90	5.60	7.00
93	52	1.08	2.07	3.10	3.62	4.13	4.65	5.17	6.20	7.23	8.27	10.33

500 Dollars.

Years.	½ per ct.	1 per ct.	2 per ct.	3 per ct.	3½ per ct.	4 per ct.	4½ per ct.	5 per ct.	6 per ct.	7 per ct.	8 per ct.	10 per ct.
1	2.50	5.00	10.00	15.00	17.50	20.00	22.50	25.00	30.00	35.00	40.00	50.00
2	5.00	10.00	20.00	30.00	35.00	40.00	45.00	50.00	60.00	70.00	80.00	100.00
3	7.50	15.00	30.00	45.00	52.50	60.00	67.50	75.00	90.00	105.00	120.00	150.00
4	10.00	20.00	40.00	60.00	70.00	80.00	90.00	100.00	120.00	140.00	160.00	200.00
5	12.50	25.00	50.00	75.00	87.50	100.00	112.50	125.00	150.00	175.00	200.00	250.00
Months.												
1	21	42	83	1.25	1.46	1.67	1.88	2.08	2.50	2.92	3.33	4.17
2	42	83	1.67	2.50	2.92	3.33	3.75	4.17	5.00	5.83	6.67	8.33
3	63	1.25	2.50	3.75	4.37	5.00	5.63	6.25	7.50	8.75	10.00	12.50
4	83	1.67	3.33	5.00	5.83	6.67	7.50	8.33	10.00	11.67	13.33	16.67
5	1.04	2.08	4.17	6.25	7.29	8.33	9.38	10.42	12.50	14.58	16.67	20.83
6	1.25	2.50	5.00	7.50	8.75	10.00	11.25	12.50	15.00	17.50	20.00	25.00
7	1.46	2.92	5.83	8.75	10.21	11.67	13.13	14.58	17.50	20.42	23.33	29.17
8	1.67	3.33	6.67	10.00	11.67	13.33	15.00	16.67	20.00	23.33	26.67	33.33
9	1.88	3.75	7.50	11.25	13.13	15.00	16.88	18.75	22.50	26.25	30.00	37.50
10	2.08	4.17	8.33	12.50	14.58	16.67	18.75	20.83	25.00	29.17	33.33	41.67
11	2.29	4.58	9.17	13.75	16.04	18.33	20.63	22.92	27.50	32.08	36.67	45.83
Days.												
1	01	01	03	04	05	06	06	07	08	10	11	14
2	01	03	06	08	10	11	13	14	17	19	22	28
3	02	04	08	13	15	17	19	21	25	29	33	42
4	03	06	11	17	19	22	25	28	33	39	44	56
5	03	07	14	21	24	28	31	35	42	49	56	69
6	04	08	17	25	29	33	38	42	50	58	67	83
7	05	10	19	29	34	39	44	49	58	68	78	97
8	06	11	22	33	39	44	50	56	67	78	89	1.11
9	06	13	25	38	44	50	56	63	75	88	1.00	1.25
10	07	14	28	42	49	56	63	69	83	97	1.11	1.39
11	08	15	31	46	53	61	69	76	92	1.07	1.22	1.53
12	08	17	33	50	58	67	75	83	1.00	1.17	1.33	1.67
13	09	18	36	54	63	72	81	90	1.08	1.26	1.44	1.81
14	10	19	39	58	68	78	88	97	1.17	1.36	1.56	1.94
15	10	21	42	63	73	83	94	1.04	1.25	1.46	1.67	2.08
16	11	22	44	67	78	89	1.00	1.11	1.33	1.56	1.78	2.22
17	12	24	47	71	83	94	1.06	1.18	1.42	1.65	1.89	2.36
18	13	25	50	75	88	1.00	1.13	1.25	1.50	1.75	2.00	2.50
19	13	26	53	79	92	1.06	1.19	1.32	1.58	1.85	2.11	2.64
20	14	28	56	83	97	1.11	1.25	1.39	1.67	1.94	2.22	2.78
21	15	29	58	88	1.02	1.17	1.31	1.46	1.75	2.04	2.33	2.92
22	15	31	61	92	1.07	1.22	1.38	1.53	1.83	2.14	2.44	3.06
23	16	32	64	96	1.12	1.28	1.44	1.60	1.92	2.24	2.56	3.19
24	17	33	67	1.00	1.17	1.33	1.50	1.67	2.00	2.33	2.67	3.33
25	17	35	69	1.04	1.22	1.39	1.56	1.74	2.08	2.43	2.78	3.47
26	18	36	72	1.08	1.26	1.44	1.63	1.81	2.17	2.53	2.89	3.61
27	19	38	75	1.13	1.31	1.50	1.69	1.88	2.25	2.63	3.00	3.75
28	19	39	78	1.17	1.36	1.56	1.75	1.94	2.33	2.72	3.11	3.89
29	20	40	81	1.21	1.41	1.61	1.81	2.01	2.42	2.82	3.22	4.03
30	21	42	83	1.25	1.46	1.67	1.88	2.08	2.50	2.92	3.33	4.17
33	23	46	92	1.38	1.60	1.83	2.06	2.29	2.75	3.21	3.66	4.58
63	44	88	1.75	2.63	3.06	3.50	3.94	4.38	5.25	6.13	7.00	8.75
93	65	1.29	2.58	3.88	4.52	5.17	5.81	6.46	7.75	9.04	10.33	12.92

600 Dollars.

Years.	½ per ct.	1 per ct.	2 per ct.	3 per ct.	3½ per ct.	4 per ct.	4½ per ct.	5 per ct.	6 per ct.	7 per ct.	8 per ct.	10 per ct.
1	3.00	6.00	12.00	18.00	21.00	24.00	27.00	30.00	36.00	42.00	48.00	60.00
2	6.00	12.00	24.00	36.00	42.00	48.00	54.00	60.00	72.00	84.00	96.00	120.00
3	9.00	18.00	36.00	54.00	63.00	72.00	81.00	90.00	108.00	126.00	144.00	180.00
4	12.00	24.00	48.00	72.00	84.00	96.00	108.00	120.00	144.00	168.00	192.00	240.00
5	15.00	30.00	60.00	90.00	105.00	120.00	135.00	150.00	180.00	210.00	240.00	300.00
Months.												
1	25	50	1.00	1.50	1.75	2.00	2.25	2.50	3.00	3.50	4.00	5.00
2	50	1.00	2.00	3.00	3.50	4.00	4.50	5.00	6.00	7.00	8.00	10.00
3	75	1.50	3.00	4.50	5.25	6.00	6.75	7.50	9.00	10.50	12.00	15.00
4	1.00	2.00	4.00	6.00	7.00	8.00	9.00	10.00	12.00	14.00	16.00	20.00
5	1.25	2.50	5.00	7.50	8.75	10.00	11.25	12.50	15.00	17.50	20.00	25.00
6	1.50	3.00	6.00	9.00	10.50	12.00	13.50	15.00	18.00	21.00	24.00	30.00
7	1.75	3.50	7.00	10.50	12.25	14.00	15.75	17.50	21.00	24.50	28.00	35.00
8	2.00	4.00	8.00	12.00	14.00	16.00	18.00	20.00	24.00	28.00	32.00	40.00
9	2.25	4.50	9.00	13.50	15.75	18.00	20.25	22.50	27.00	31.50	36.00	45.00
10	2.50	5.00	10.00	15.00	17.50	20.00	22.50	25.00	30.00	35.00	40.00	50.00
11	2.75	5.50	11.00	16.50	19.25	22.00	24.75	27.50	33.00	38.50	44.00	55.00
Days.												
1	01	02	03	05	06	07	08	08	10	12	13	17
2	02	03	07	10	12	13	15	17	20	23	27	33
3	03	05	10	15	18	20	23	25	30	35	40	50
4	03	07	13	20	23	27	30	33	40	47	53	67
5	04	08	17	25	29	33	38	42	50	58	67	83
6	05	10	20	30	35	40	45	50	60	70	80	1.00
7	06	12	23	35	41	47	53	58	70	82	93	1.17
8	07	13	27	40	47	53	60	67	80	93	1.07	1.33
9	08	15	30	45	53	60	68	75	90	1.05	1.20	1.50
10	08	17	33	50	58	67	75	83	1.00	1.17	1.33	1.67
11	09	18	37	55	64	73	83	92	1.10	1.28	1.47	1.83
12	10	20	40	60	70	80	90	1.00	1.20	1.40	1.60	2.00
13	11	22	43	65	76	87	98	1.08	1.30	1.52	1.73	2.17
14	12	23	47	70	82	93	1.05	1.17	1.40	1.63	1.87	2.33
15	13	25	50	75	88	1.00	1.13	1.25	1.50	1.75	2.00	2.50
16	13	27	53	80	93	1.07	1.20	1.33	1.60	1.87	2.13	2.67
17	14	28	57	85	99	1.13	1.28	1.42	1.70	1.98	2.27	2.83
18	15	30	60	90	1.05	1.20	1.35	1.50	1.80	2.10	2.40	3.00
19	16	32	63	95	1.11	1.27	1.43	1.58	1.90	2.22	2.53	3.17
20	17	33	67	1.00	1.17	1.33	1.50	1.67	2.00	2.33	2.67	3.33
21	18	35	70	1.05	1.23	1.40	1.58	1.75	2.10	2.45	2.80	3.50
22	18	37	73	1.10	1.28	1.47	1.65	1.83	2.20	2.57	2.93	3.67
23	19	38	77	1.15	1.34	1.53	1.73	1.92	2.30	2.68	3.07	3.83
24	20	40	80	1.20	1.40	1.60	1.80	2.00	2.40	2.80	3.20	4.00
25	21	42	83	1.25	1.46	1.67	1.88	2.08	2.50	2.92	3.33	4.17
26	22	43	87	1.30	1.52	1.73	1.95	2.17	2.60	3.03	3.47	4.33
27	23	45	90	1.35	1.58	1.80	2.03	2.25	2.70	3.15	3.60	4.50
28	23	47	93	1.40	1.63	1.87	2.10	2.33	2.80	3.27	3.73	4.67
29	24	48	97	1.45	1.69	1.93	2.18	2.42	2.90	3.38	3.87	4.83
30	25	50	1.00	1.50	1.75	2.00	2.25	2.50	3.00	3.50	4.00	5.00
33	28	55	1.10	1.65	1.93	2.20	2.48	2.75	3.30	3.85	4.40	5.50
63	53	1.05	2.10	3.15	3.68	4.20	4.73	5.25	6.30	7.35	8.40	10.50
93	77	1.55	3.10	4.65	5.43	6.20	6.98	7.75	9.30	10.85	12.40	15.50

700 Dollars.

Years.	½ per ct.	1 per ct.	2 per ct.	3 per ct.	3½ per ct.	4 per ct.	4½ per ct.	5 per ct.	6 per ct.	7 per ct.	8 per ct.	10 per ct.
1	3.50	7.00	14.00	21.00	24.50	28.00	31.50	35.00	42.00	49.00	56.00	70.00
2	7.00	14.00	28.00	42.00	49.00	56.00	63.00	70.00	84.00	98.00	112.00	140.00
3	10.50	21.00	42.00	63.00	73.50	84.00	94.50	105.00	126.00	147.00	168.00	210.00
4	14.00	28.00	56.00	84.00	98.00	112.00	126.00	140.00	168.00	196.00	224.00	280.00
5	17.50	35.00	70.00	105.00	122.50	140.00	157.50	175.00	210.00	245.00	280.00	350.00

Months.												
1	29	58	1.17	1.75	2.04	2.33	2.63	2.90	3.50	4.08	4.67	5.83
2	58	1.17	2.33	3.50	4.08	4.67	5.25	5.83	7.00	8.17	9.33	11.67
3	88	1.75	3.50	5.25	6.12	7.00	7.88	8.75	10.50	12.25	14.00	17.50
4	1.17	2.33	4.67	7.00	8.17	9.33	10.50	11.67	14.00	16.33	18.67	23.33
5	1.46	2.92	5.83	8.75	10.21	11.67	13.13	14.58	17.50	20.42	23.33	29.17
6	1.75	3.50	7.00	10.50	12.25	14.00	15.75	17.50	21.00	24.50	28.00	35.00
7	2.04	4.08	8.17	12.25	14.29	16.33	18.38	20.42	24.50	28.58	32.67	40.83
8	2.33	4.67	9.33	14.00	16.33	18.67	21.00	23.33	28.00	32.67	37.33	46.67
9	2.63	5.25	10.50	15.75	18.38	21.00	23.63	26.25	31.50	36.75	42.00	52.50
10	2.92	5.83	11.67	17.50	20.42	23.33	26.25	29.17	35.00	40.83	46.67	58.33
11	3.21	6.42	12.83	19.25	22.46	25.67	28.88	32.08	38.50	44.92	51.33	64.17

Days.												
1	01	02	04	06	07	08	09	10	12	14	16	19
2	02	04	08	12	14	16	18	19	23	27	31	39
3	03	06	12	18	20	23	26	29	35	41	47	58
4	04	08	16	23	27	31	35	39	47	54	62	78
5	05	10	19	29	34	39	44	49	58	68	78	97
6	06	12	23	35	41	47	53	58	70	82	93	1.17
7	07	14	27	41	48	54	61	68	82	95	1.09	1.36
8	08	16	31	47	54	62	70	78	93	1.09	1.24	1.56
9	09	18	35	53	61	70	79	88	1.05	1.23	1.40	1.75
10	10	19	39	58	68	78	88	97	1.17	1.36	1.56	1.94
11	11	21	43	64	75	86	96	1.07	1.28	1.50	1.71	2.14
12	12	23	47	70	82	93	1.05	1.17	1.40	1.63	1.87	2.33
13	13	25	51	76	88	1.01	1.14	1.26	1.52	1.77	2.02	2.53
14	14	27	54	82	95	1.09	1.23	1.36	1.63	1.91	2.18	2.72
15	15	29	58	88	1.02	1.17	1.31	1.46	1.75	2.04	2.33	2.92
16	16	31	62	93	1.09	1.24	1.40	1.56	1.87	2.18	2.49	3.11
17	17	33	66	99	1.16	1.32	1.49	1.65	1.98	2.31	2.64	3.31
18	18	35	70	1.05	1.23	1.40	1.58	1.75	2.10	2.45	2.80	3.50
19	18	37	74	1.11	1.29	1.48	1.66	1.85	2.22	2.59	2.96	3.69
20	19	39	78	1.17	1.36	1.56	1.75	1.94	2.33	2.72	3.11	3.89
21	20	41	82	1.23	1.43	1.63	1.84	2.04	2.45	2.86	3.27	4.08
22	21	43	86	1.28	1.50	1.71	1.93	2.14	2.57	2.99	3.42	4.28
23	22	45	89	1.34	1.57	1.79	2.01	2.24	2.68	3.13	3.58	4.47
24	23	47	93	1.40	1.63	1.87	2.10	2.33	2.80	3.27	3.73	4.67
25	24	49	97	1.46	1.70	1.94	2.19	2.43	2.92	3.40	3.89	4.86
26	25	51	1.01	1.52	1.77	2.02	2.28	2.53	3.03	3.54	4.04	5.06
27	26	53	1.05	1.58	1.84	2.10	2.36	2.63	3.15	3.68	4.20	5.25
28	27	54	1.09	1.63	1.91	2.18	2.45	2.72	3.27	3.81	4.36	5.44
29	28	56	1.13	1.69	1.97	2.26	2.54	2.82	3.38	3.95	4.51	5.64
30	29	58	1.17	1.75	2.04	2.33	2.63	2.92	3.50	4.08	4.67	5.83
33	32	64	1.28	1.93	2.25	2.57	2.89	3.21	3.85	4.49	5.13	6.42
63	61	1.23	2.45	3.68	4.30	4.90	5.51	6.13	7.35	8.58	9.80	12.25
93	90	1.81	3.62	5.43	6.33	7.23	8.14	9.04	10.85	12.66	14.47	18.08

800 Dollars.

Years.	½ per ct.	1 per ct.	2 per ct.	3 per ct.	3½ per ct.	4 per ct.	4½ per ct.	5 per ct.	6 per ct.	7 per ct.	8 per ct.	10 per ct.
1	4.00	8.00	16.00	24.00	28.00	32.00	36.00	40.00	48.00	56.00	64.00	80.00
2	8.00	16.00	32.00	48.00	56.00	64.00	72.00	80.00	96.00	112.00	128.00	160.00
3	12.00	24.00	48.00	72.00	84.00	96.00	108.00	120.00	144.00	168.00	192.00	240.00
4	16.00	32.00	64.00	96.00	112.00	128.00	144.00	160.00	192.00	224.00	256.00	320.00
5	20.00	40.00	80.00	120.00	140.00	160.00	180.00	200.00	240.00	280.00	320.00	400.00

Months.

	½ per ct.	1 per ct.	2 per ct.	3 per ct.	3½ per ct.	4 per ct.	4½ per ct.	5 per ct.	6 per ct.	7 per ct.	8 per ct.	10 per ct.
1	33	67	1.33	2.00	2.33	2.67	3.00	3.33	4.00	4.67	5.33	6.67
2	67	1.33	2.67	4.00	4.67	5.33	6.00	6.67	8.00	9.33	10.67	13.33
3	1.00	2.00	4.00	6.00	7.00	8.00	9.00	10.00	12.00	14.00	16.00	20.00
4	1.33	2.67	5.33	8.00	9.33	10.67	12.00	13.33	16.00	18.67	21.33	26.67
5	1.67	3.33	6.67	10.00	11.67	13.33	15.00	16.67	20.00	23.33	26.67	33.33
6	2.00	4.00	8.00	12.00	14.00	16.00	18.00	20.00	24.00	28.00	32.00	40.00
7	2.33	4.67	9.33	14.00	16.33	18.67	21.00	23.33	28.00	32.67	37.33	46.67
8	2.67	5.33	10.67	16.00	18.67	21.33	24.00	26.67	32.00	37.33	42.67	53.33
9	3.00	6.00	12.00	18.00	21.00	24.00	27.00	30.00	36.00	42.00	48.00	60.00
10	3.33	6.67	13.33	20.00	23.33	26.67	30.00	33.33	40.00	46.67	53.33	66.67
11	3.67	7.33	14.67	22.00	25.67	29.33	33.00	36.67	44.00	51.33	58.67	73.33

Days.

	½ per ct.	1 per ct.	2 per ct.	3 per ct.	3½ per ct.	4 per ct.	4½ per ct.	5 per ct.	6 per ct.	7 per ct.	8 per ct.	10 per ct.
1	01	02	04	07	08	09	10	11	13	16	18	22
2	02	04	09	13	16	18	20	22	27	31	36	44
3	03	07	13	20	23	27	30	33	40	47	53	67
4	04	09	18	27	31	36	40	44	53	62	71	89
5	06	11	22	33	39	44	50	56	67	78	89	1.11
6	07	13	27	40	47	53	60	67	80	93	1.07	1.33
7	08	16	31	47	54	62	70	78	93	1.09	1.24	1.56
8	09	18	36	53	62	71	80	89	1.07	1.24	1.42	1.78
9	10	20	40	60	70	80	90	1.00	1.20	1.40	1.60	2.00
10	11	22	44	67	78	89	1.00	1.11	1.33	1.56	1.78	2.22
11	12	24	49	73	86	98	1.10	1.22	1.47	1.71	1.96	2.44
12	13	27	53	80	93	1.07	1.20	1.33	1.60	1.87	2.13	2.67
13	14	29	58	87	1.01	1.16	1.30	1.44	1.73	2.02	2.31	2.89
14	16	31	62	93	1.09	1.24	1.40	1.56	1.87	2.18	2.49	3.11
15	17	33	67	1.00	1.17	1.33	1.50	1.67	2.00	2.33	2.67	3.33
16	18	36	71	1.07	1.24	1.42	1.60	1.78	2.13	2.49	2.84	3.56
17	19	38	76	1.13	1.32	1.51	1.70	1.89	2.27	2.64	3.02	3.78
18	20	40	80	1.20	1.40	1.60	1.80	2.00	2.40	2.80	3.20	4.00
19	21	42	84	1.27	1.48	1.69	1.90	2.11	2.53	2.96	3.38	4.22
20	22	44	89	1.33	1.56	1.78	2.00	2.22	2.67	3.11	3.56	4.44
21	23	47	93	1.40	1.63	1.87	2.10	2.33	2.80	3.27	3.73	4.67
22	24	49	98	1.47	1.71	1.96	2.20	2.44	2.93	3.42	3.91	4.89
23	26	51	1.02	1.53	1.79	2.04	2.30	2.56	3.07	3.58	4.09	5.11
24	27	53	1.07	1.60	1.87	2.13	2.40	2.67	3.20	3.73	4.27	5.33
25	28	56	1.11	1.67	1.94	2.22	2.50	2.78	3.33	3.89	4.44	5.56
26	29	58	1.16	1.73	2.02	2.31	2.60	2.89	3.47	4.04	4.62	5.78
27	30	60	1.20	1.80	2.10	2.40	2.70	3.00	3.60	4.20	4.80	6.00
28	31	62	1.24	1.87	2.18	2.49	2.80	3.11	3.73	4.36	4.98	6.22
29	32	64	1.29	1.93	2.26	2.58	2.90	3.22	3.87	4.51	5.16	6.44
30	33	67	1.33	2.00	2.33	2.67	3.00	3.33	4.00	4.67	5.33	6.67
33	37	73	1.46	2.20	2.57	2.93	3.30	3.67	4.40	5.13	5.87	7.33
63	70	1.40	2.80	4.20	4.90	5.60	6.30	7.00	8.40	9.80	11.20	14.00
93	1.03	2.07	4.13	6.20	7.23	8.27	9.30	10.33	12.40	14.47	16.53	20.67

900 Dollars.

Years.	½ per ct.	1 per ct.	2 per ct.	3 per ct.	3½ per ct.	4 per ct.	4½ per ct.	5 per ct.	6 per ct.	7 per ct.	8 per ct.	10 per ct.
1	4.50	9.00	18.00	27.00	31.50	36.00	40.50	45.00	54.00	63.00	72.00	90.00
2	9.00	18.00	36.00	54.00	63.00	72.00	81.00	90.00	108.00	126.00	144.00	180.00
3	13.50	27.00	54.00	81.00	94.50	108.00	121.50	135.00	162.00	189.00	216.00	270.00
4	18.00	36.00	72.00	108.00	126.00	144.00	162.00	180.00	216.00	252.00	288.00	360.00
5	22.50	45.00	90.00	135.00	157.50	180.00	202.50	225.00	270.00	315.00	360.00	450.00

Months.												
1	38	75	1.50	2.25	2.63	3.00	3.38	3.75	4.50	5.25	6.00	7.50
2	75	1.50	3.00	4.50	5.25	6.00	6.75	7.50	9.00	10.50	12.00	15.00
3	1.13	2.25	4.50	6.75	7.88	9.00	10.13	11.25	13.50	15.75	18.00	22.50
4	1.50	3.00	6.00	9.00	10.50	12.00	13.50	15.00	18.00	21.00	24.00	30.00
5	1.88	3.75	7.50	11.25	13.13	15.00	16.88	18.75	22.50	26.25	30.00	37.50
6	2.25	4.50	9.00	13.50	15.75	18.00	20.25	22.50	27.00	31.50	36.00	45.00
7	2.62	5.25	10.50	15.75	18.38	21.00	23.63	26.25	31.50	36.75	42.00	52.50
8	3.00	6.00	12.00	18.00	21.00	24.00	27.00	30.00	36.00	42.00	48.00	60.00
9	3.38	6.75	13.50	20.25	23.63	27.00	30.38	33.75	40.50	47.25	54.00	67.50
10	3.75	7.50	15.00	22.50	26.25	30.00	33.75	37.50	45.00	52.50	60.00	75.00
.11	4.13	8.25	16.50	24.75	28.88	33.00	37.13	41.25	49.50	57.75	66.00	82.50

Days.												
1	01	03	05	08	09	10	11	13	15	18	20	25
2	03	05	10	15	18	20	23	25	30	35	40	50
3	04	08	15	23	26	30	34	38	45	53	60	75
4	05	10	20	30	35	40	45	50	60	70	80	1.00
5	06	13	25	38	44	50	56	63	75	88	1.00	1.25
6	08	15	30	45	53	60	68	75	90	1.05	1.20	1.50
7	09	18	35	53	61	70	79	88	1.05	1.23	1.40	1.75
8	10	20	40	60	70	80	90	1.00	1.20	1.40	1.60	2.00
9	11	23	45	68	79	90	1.01	1.13	1.35	1.58	1.80	2.25
10	13	25	50	75	88	1.00	1.13	1.25	1.50	1.75	2.00	2.50
11	14	28	55	83	96	1.10	1.24	1.38	1.65	1.93	2.20	2.75
12	15	30	60	90	1.05	1.20	1.35	1.50	1.80	2.10	2.40	3.00
13	16	33	65	98	1.14	1.30	1.46	1.63	1.95	2.28	2.60	3.25
14	18	35	70	1.05	1.25	1.40	1.58	1.75	2.10	2.45	2.80	3.50
15	19	38	75	1.13	1.31	1.50	1.69	1.88	2.25	2.63	3.00	3.75
16	20	40	80	1.20	1.40	1.60	1.80	2.00	2.40	2.80	3.20	4.00
17	21	43	85	1.28	1.49	1.70	1.91	2.13	2.55	2.98	3.40	4.25
18	23	45	90	1.35	1.58	1.80	2.03	2.25	2.70	3.15	3.60	4.50
19	24	48	95	1.43	1.66	1.90	2.14	2.38	2.85	3.33	3.80	4.75
20	25	50	1.00	1.50	1.75	2.00	2.25	2.50	3.00	3.50	4.00	5.00
21	26	53	1.05	1.58	1.84	2.10	2.36	2.63	3.15	3.68	4.20	5.25
22	28	55	1.10	1.65	1.93	2.20	2.48	2.75	3.30	3.85	4.40	5.50
23	29	58	1.15	1.73	2.01	2.30	2.59	2.88	3.45	4.03	4.60	5.75
24	30	60	1.20	1.80	2.10	2.40	2.70	3.00	3.60	4.20	4.80	6.00
25	31	63	1.25	1.88	2.19	2.50	2.81	3.13	3.75	4.38	5.00	6.25
26	33	65	1.30	1.95	2.28	2.60	2.93	3.25	3.90	4.55	5.20	6.50
27	34	68	1.35	2.03	2.36	2.70	3.04	3.38	4.05	4.73	5.40	6.75
28	35	70	1.40	2.10	2.45	2.80	3.15	3.50	4.20	4.90	5.60	7.00
29	36	73	1.45	2.18	2.54	2.90	3.26	3.63	4.35	5.08	5.80	7.25
30	38	75	1.50	2.25	2.63	3.00	3.38	3.75	4.50	5.25	6.00	7.50
33	41	83	1.65	2.48	2.89	3.30	3.71	4.13	4.95	5.78	6.60	8.25
63	79	1.58	3.15	4.73	5.51	6.30	7.09	7.88	9.45	11.03	12.60	15.75
93	1.16	2.33	4.65	6.98	8.14	9.30	10.46	11.63	13.95	16.28	18.60	23.25

1000 Dollars.

Years.	½ per ct.	1 per ct.	2 per ct.	3 per ct.	3½ per ct.	4 per ct.	4½ per ct.	5 per ct.	6 per ct.	7 per ct.	8 per ct.	10 per ct.
1	5.00	10.00	20.00	30.00	35.00	40.00	45.00	50.00	60.00	70.00	80.00	100.00
2	10.00	20.00	40.00	60.00	70.00	80.00	90.00	100.00	120.00	140.00	160.00	200.00
3	15.00	30.00	60.00	90.00	105.00	120.00	135.00	150.00	180.00	210.00	240.00	300.00
4	20.00	40.00	80.00	120.00	140.00	160.00	180.00	200.00	240.00	280.00	320.00	400.00
5	25.00	50.00	100.00	150.00	175.00	200.00	225.00	250.00	300.00	350.00	400.00	500.00
Months.												
1	42	83	1.67	2.50	2.92	3.33	3.75	4.17	5.00	5.83	6.67	8.33
2	83	1.67	3.33	5.00	5.83	6.67	7.50	8.33	10.00	11.67	13.33	16.67
3	1.25	2.50	5.00	7.50	8.75	10.00	11.25	12.50	15.00	17.50	20.00	25.00
4	1.67	3.33	6.67	10.00	11.67	13.33	15.00	16.67	20.00	23.33	26.67	33.33
5	2.08	4.17	8.33	12.50	14.58	16.67	18.75	20.83	25.00	29.17	33.33	41.67
6	2.50	5.00	10.00	15.00	17.50	20.00	22.50	25.00	30.00	35.00	40.00	50.00
7	2.92	5.83	11.67	17.50	20.42	23.33	26.25	29.17	35.00	40.83	46.67	58.33
8	3.33	6.67	13.33	20.00	23.33	26.67	30.00	33.33	40.00	46.67	53.33	66.67
9	3.75	7.50	15.00	22.50	26.25	30.00	33.75	37.50	45.00	52.50	60.00	75.00
10	4.17	8.33	16.67	25.00	29.17	33.33	37.50	41.67	50.00	58.33	66.67	83.33
11	4.58	9.17	18.33	27.50	32.08	36.67	41.25	45.83	55.00	64.17	73.33	91.67
Days.												
1	01	03	06	08	10	11	13	14	17	19	22	28
2	03	06	11	17	19	22	25	28	33	39	44	56
3	04	08	17	25	29	33	38	42	50	58	67	83
4	06	11	22	33	39	44	50	56	67	78	89	1.11
5	07	14	28	42	49	56	63	69	83	97	1.11	1.39
6	08	17	33	50	58	67	75	83	1.00	1.17	1.33	1.67
7	10	19	39	58	68	78	88	97	1.17	1.36	1.56	1.94
8	11	22	44	67	78	89	1.00	1.11	1.33	1.56	1.78	2.22
9	13	25	50	75	88	1.00	1.13	1.25	1.50	1.75	2.00	2.50
10	14	28	56	83	97	1.11	1.25	1.39	1.67	1.94	2.22	2.78
11	15	31	61	92	1.07	1.22	1.38	1.53	1.83	2.14	2.44	3.06
12	17	33	67	1.00	1.17	1.33	1.50	1.67	2.00	2.33	2.67	3.33
13	18	36	72	1.08	1.26	1.44	1.63	1.81	2.17	2.53	2.89	3.61
14	19	39	78	1.17	1.36	1.56	1.75	1.94	2.33	2.72	3.11	3.89
15	21	42	83	1.25	1.46	1.67	1.88	2.08	2.50	2.92	3.33	4.17
16	22	44	89	1.33	1.56	1.78	2.00	2.22	2.67	3.11	3.56	4.44
17	24	47	94	1.42	1.65	1.89	2.13	2.36	2.83	3.31	3.78	4.72
18	25	50	1.00	1.50	1.75	2.00	2.25	2.50	3.00	3.50	4.00	5.00
19	26	53	1.06	1.58	1.85	2.11	2.38	2.64	3.17	3.69	4.22	5.28
20	28	56	1.11	1.67	1.94	2.22	2.50	2.78	3.33	3.89	4.44	5.56
21	29	58	1.17	1.75	2.04	2.33	2.63	2.92	3.50	4.08	4.67	5.83
22	31	61	1.22	1.83	2.14	2.44	2.75	3.06	3.67	4.28	4.89	6.11
23	32	64	1.28	1.92	2.24	2.56	2.88	3.19	3.83	4.47	5.11	6.39
24	33	67	1.33	2.00	2.33	2.67	3.00	3.33	4.00	4.67	5.33	6.67
25	35	69	1.39	2.08	2.43	2.78	3.13	3.47	4.17	4.86	5.56	6.94
26	36	72	1.44	2.17	2.53	2.89	3.25	3.61	4.33	5.06	5.78	7.22
27	38	75	1.50	2.25	2.63	3.00	3.38	3.75	4.50	5.25	6.00	7.50
28	39	78	1.56	2.33	2.72	3.11	3.50	3.89	4.67	5.44	6.22	7.78
29	40	81	1.61	2.42	2.82	3.22	3.63	4.03	4.83	5.64	6.44	8.06
30	42	83	1.67	2.50	2.92	3.33	3.75	4.17	5.00	5.83	6.67	8.33
33	46	91	1.83	2.75	3.21	3.67	4.13	4.58	5.50	6.42	7.33	9.17
63	88	1.75	3.50	5.25	6.13	7.00	7.88	8.75	10.50	12.25	14.00	17.50
93	1.29	2.58	5.17	7.75	9.04	10.33	11.63	12.92	15.50	18.08	20.67	25.83

2000 Dollars.

	½ per ct.	1 per ct.	2 per ct.	3 per ct.	3½ per ct.	4 per ct.	4½ per ct.	5 per ct.	6 per ct.	7 per ct.	8 per ct.	10 per ct.
Years.												
1	10.00	20.00	40.00	60.00	70.00	80.00	90.00	100.00	120.00	140.00	160.00	200.00
2	20.00	40.00	80.00	120.00	140.00	160.00	180.00	200.00	240.00	280.00	320.00	400.00
3	30.00	60.00	120.00	180.00	210.00	240.00	270.00	300.00	360.00	420.00	480.00	600.00
4	40.00	80.00	160.00	240.00	280.00	320.00	360.00	400.00	480.00	560.00	640.00	800.00
5	50.00	100.00	200.00	300.00	350.00	400.00	450.00	500.00	600.00	700.00	800.00	1000.00
Months.												
1	83	1.67	3.33	5.00	5.83	6.67	7.50	8.33	10.00	11.67	13.33	16.67
2	1.67	3.33	6.67	10.00	11.67	13.33	15.00	16.67	20.00	23.33	26.67	33.33
3	2.50	5.00	10.00	15.00	17.50	20.00	22.50	25.00	30.00	35.00	40.00	50.00
4	3.33	6.67	13.33	20.00	23.33	26.67	30.00	33.33	40.00	46.67	53.33	66.67
5	4.17	8.33	16.67	25.00	29.17	33.33	37.50	41.67	50.00	58.33	66.67	83.33
6	5.00	10.00	20.00	30.00	35.00	40.00	45.00	50.00	60.00	70.00	80.00	100.00
7	5.83	11.67	23.33	35.00	40.83	46.67	52.50	58.33	70.00	81.67	93.33	116.67
8	6.67	13.33	26.67	40.00	46.67	53.33	60.00	66.67	80.00	93.33	106.67	133.33
9	7.50	15.00	30.00	45.00	52.50	60.00	67.50	75.00	90.00	105.00	120.00	150.00
10	8.33	16.67	33.33	50.00	58.33	66.67	75.00	83.33	100.00	116.67	133.33	166.67
11	9.17	18.33	36.67	55.00	64.17	73.33	82.50	91.67	110.00	128.33	146.67	183.33
Days.												
1	03	06	11	17	19	22	25	28	33	39	44	56
2	06	11	22	33	39	44	50	56	67	78	89	1.11
3	08	17	33	50	58	67	75	83	1.00	1.17	1.33	1.67
4	11	22	44	67	78	89	1.00	1.11	1.33	1.56	1.78	2.22
5	14	28	56	83	97	1.11	1.25	1.39	1.67	1.94	2.22	2.78
6	17	33	67	1.00	1.17	1.33	1.50	1.67	2.00	2.33	2.67	3.33
7	19	39	78	1.17	1.36	1.56	1.75	1.94	2.33	2.72	3.11	3.89
8	22	44	89	1.33	1.56	1.78	2.00	2.22	2.67	3.11	3.56	4.44
9	25	50	1.00	1.50	1.75	2.00	2.25	2.50	3.00	3.50	4.00	5.00
10	28	56	1.11	1.67	1.94	2.22	2.50	2.78	3.33	3.89	4.44	5.56
11	31	61	1.22	1.83	2.14	2.44	2.75	3.06	3.67	4.28	4.89	6.11
12	33	67	1.33	2.00	2.33	2.67	3.00	3.33	4.00	4.67	5.33	6.67
13	36	72	1.44	2.17	2.53	2.89	3.25	3.61	4.33	5.06	5.78	7.22
14	39	78	1.56	2.33	2.72	3.11	3.50	3.89	4.67	5.44	6.22	7.78
15	42	83	1.67	2.50	2.92	3.33	3.75	4.17	5.00	5.83	6.67	8.33
16	44	89	1.78	2.67	3.11	3.56	4.00	4.44	5.33	6.22	7.11	8.89
17	47	94	1.89	2.83	3.31	3.78	4.25	4.72	5.67	6.61	7.56	9.44
18	50	1.00	2.00	3.00	3.50	4.00	4.50	5.00	6.00	7.00	8.00	10.00
19	53	1.06	2.11	3.17	3.69	4.22	4.75	5.28	6.33	7.39	8.44	10.56
20	56	1.11	2.22	3.33	3.89	4.44	5.00	5.56	6.67	7.78	8.89	11.11
21	58	1.17	2.33	3.50	4.08	4.67	5.25	5.83	7.00	8.17	9.33	11.67
22	61	1.22	2.44	3.67	4.28	4.89	5.50	6.11	7.33	8.56	9.78	12.22
23	64	1.28	2.56	3.83	4.47	5.11	5.75	6.39	7.67	8.94	10.22	12.78
24	67	1.33	2.67	4.00	4.67	5.33	6.00	6.67	8.00	9.33	10.67	13.33
25	69	1.39	2.78	4.17	4.86	5.56	6.25	6.94	8.33	9.72	11.11	13.89
26	72	1.44	2.89	4.33	5.06	5.78	6.50	7.22	8.67	10.11	11.56	14.44
27	75	1.50	3.00	4.50	5.25	6.00	6.75	7.50	9.00	10.50	12.00	15.00
28	78	1.56	3.11	4.67	5.44	6.22	7.00	7.78	9.33	10.89	12.44	15.56
29	81	1.61	3.22	4.83	5.64	6.44	7.25	8.06	9.67	11.28	12.89	16.11
30	83	1.67	3.33	5.00	5.83	6.67	7.50	8.33	10.00	11.67	13.33	16.67
33	92	1.83	3.67	5.50	6.42	7.34	8.25	9.17	11.00	12.83	14.67	18.33
63	1.75	3.50	7.00	10.50	12.25	14.00	15.75	17.50	21.00	24.50	28.00	35.00
93	2.58	5.17	10.33	15.50	18.08	20.67	23.25	25.83	31.00	36.17	41.33	51.67

3000 Dollars.

Years.	½ per ct.	1 per ct.	2 per ct.	3 per ct.	3½ per ct.	4 per ct.	4½ per ct.	5 per ct.	6 per ct.	7 per ct.	8 per ct.	10 per ct.
1	15.00	30.00	60.00	90.00	105.00	120.00	135.00	150.00	180.00	210.00	240.00	300.00
2	30.00	60.00	120.00	180.00	210.00	240.00	270.00	300.00	360.00	420.00	480.00	600.00
3	45.00	90.00	180.00	270.00	315.00	360.00	405.00	450.00	540.00	630.00	720.00	900.00
4	60.00	120.00	240.00	360.00	420.00	480.00	540.00	600.00	720.00	840.00	960.00	1200.00
5	75.00	150.00	300.00	450.00	525.00	600.00	675.00	750.00	900.00	1050.00	1200.00	1500.00
Months.												
1	1.25	2.50	5.00	7.50	8.75	10.00	11.25	12.50	15.00	17.50	20.00	25.00
2	2.50	5.00	10.00	15.00	17.50	20.00	22.50	25.00	30.00	35.00	40.00	50.00
3	3.75	7.50	15.00	22.50	26.25	30.00	33.75	37.50	45.00	52.50	60.00	75.00
4	5.00	10.00	20.00	30.00	35.00	40.00	45.00	50.00	60.00	70.00	80.00	100.00
5	6.25	12.50	25.00	37.50	43.75	50.00	56.25	62.50	75.00	87.50	100.00	125.00
6	7.50	15.00	30.00	45.00	52.50	60.00	67.50	75.00	90.00	105.00	120.00	150.00
7	8.75	17.50	35.00	52.50	61.25	70.00	78.75	87.50	105.00	122.50	140.00	175.00
8	10.00	20.00	40.00	60.00	70.00	80.00	90.00	100.00	120.00	140.00	160.00	200.00
9	11.25	22.50	45.00	67.50	78.75	90.00	101.25	112.50	135.00	157.50	180.00	225.00
10	12.50	25.00	50.00	75.00	87.50	100.00	112.50	125.00	150.00	175.00	200.00	250.00
11	13.75	27.50	55.00	82.50	96.25	110.00	123.75	137.50	165.00	192.50	220.00	275.00
Days.												
1	04	08	17	25	29	33	38	42	50	58	67	83
2	08	17	33	50	58	67	75	83	1.00	1.17	1.33	1.67
3	13	25	50	75	88	1.00	1.13	1.25	1.50	1.75	2.00	2.50
4	17	33	67	1.00	1.17	1.33	1.50	1.67	2.00	2.33	2.67	3.33
5	21	42	83	1.25	1.46	1.67	1.88	2.08	2.50	2.92	3.33	4.17
6	25	50	1.00	1.50	1.75	2.00	2.25	2.50	3.00	3.50	4.00	5.00
7	29	58	1.17	1.75	2.04	2.33	2.63	2.92	3.50	4.08	4.67	5.83
8	33	67	1.33	2.00	2.33	2.67	3.00	3.33	4.00	4.67	5.33	6.67
9	38	75	1.50	2.25	2.63	3.00	3.38	3.75	4.50	5.25	6.00	7.50
10	42	83	1.67	2.50	2.92	3.33	3.75	4.17	5.00	5.83	6.67	8.33
11	46	92	1.83	2.75	3.21	3.67	4.13	4.58	5.50	6.42	7.33	9.17
12	50	1.00	2.00	3.00	3.50	4.00	4.50	5.00	6.00	7.00	8.00	10.00
13	54	1.08	2.17	3.25	3.79	4.33	4.88	5.42	6.50	7.58	8.67	10.83
14	58	1.17	2.33	3.50	4.08	4.67	5.25	5.83	7.00	8.17	9.33	11.67
15	63	1.25	2.50	3.75	4.38	5.00	5.62	6.25	7.50	8.75	10.00	12.50
16	67	1.33	2.67	4.00	4.67	5.33	6.00	6.67	8.00	9.33	10.67	13.33
17	71	1.42	2.83	4.25	4.95	5.67	6.38	7.08	8.50	9.92	11.33	14.17
18	75	1.50	3.00	4.50	5.25	6.00	6.75	7.50	9.00	10.50	12.00	15.00
19	79	1.58	3.17	4.75	5.54	6.33	7.13	7.92	9.50	11.08	12.67	15.83
20	83	1.67	3.33	5.00	5.83	6.67	7.50	8.33	10.00	11.67	13.33	16.67
21	88	1.75	3.50	5.25	6.13	7.00	7.88	8.75	10.50	12.25	14.00	17.50
22	92	1.83	3.67	5.50	6.42	7.33	8.25	9.17	11.00	12.83	14.67	18.33
23	96	1.92	3.83	5.75	6.71	7.67	8.63	9.58	11.50	13.42	15.33	19.17
24	1.00	2.00	4.00	6.00	7.00	8.00	9.00	10.00	12.00	14.00	16.00	20.00
25	1.04	2.08	4.17	6.25	7.29	8.33	9.38	10.42	12.50	14.58	16.67	20.83
26	1.08	2.17	4.33	6.50	7.58	8.67	9.75	10.83	13.00	15.17	17.33	21.67
27	1.13	2.25	4.50	6.75	7.88	9.00	10.13	11.25	13.50	15.75	18.00	22.50
28	1.17	2.33	4.67	7.00	8.17	9.33	10.50	11.67	14.00	16.33	18.67	23.33
29	1.21	2.42	4.83	7.25	8.46	9.67	10.88	12.08	14.50	16.92	19.33	24.17
30	1.25	2.50	5.00	7.50	8.75	10.00	11.25	12.50	15.00	17.50	20.00	25.00
33	1.38	2.75	5.50	8.25	9.63	11.00	12.38	13.75	16.50	19.25	22.00	27.50
63	2.63	5.25	10.50	15.75	18.38	21.00	23.63	26.25	31.50	36.75	42.00	52.50
93	3.88	7.75	15.50	23.25	27.13	31.00	34.88	38.75	46.50	54.25	62.00	77.50

4000 Dollars.

	½ per ct.	1 per ct.	2 per ct.	3 per ct.	3½ per ct.	4 per ct.	4½ per ct.	5 per ct.	6 per ct.	7 per ct.	8 per ct.	10 per ct.
Years.												
1	20.00	40.00	80.00	120.00	140.00	160.00	180.00	200.00	240.00	280.00	320.00	400.00
2	40.00	80.00	160.00	240.00	280.00	320.00	360.00	400.00	480.00	560.00	640.00	800.00
3	60.00	120.00	240.00	360.00	420.00	480.00	540.00	600.00	720.00	840.00	960.00	1200.00
4	80.00	160.00	320.00	480.00	560.00	640.00	720.00	800.00	960.00	1120.00	1280.00	1600.00
5	100.00	200.00	400.00	600.00	700.00	800.00	900.00	1000.00	1200.00	1400.00	1600.00	2000.00
Months.												
1	1.67	3.33	6.67	10.00	11.67	13.33	15.00	16.67	20.00	23.33	26.67	33.33
2	3.33	6.67	13.33	20.00	23.33	26.67	30.00	33.33	40.00	46.67	53.33	66.67
3	5.00	10.00	20.00	30.00	35.00	40.00	45.00	50.00	60.00	70.00	80.00	100.00
4	6.67	13.33	26.67	40.00	46.67	53.33	60.00	66.67	80.00	93.33	106.67	133.33
5	8.33	16.67	33.33	50.00	58.33	66.67	75.00	83.33	100.00	116.67	133.33	166.67
6	10.00	20.00	40.00	60.00	70.00	80.00	90.00	100.00	120.00	140.00	160.00	200.00
7	11.67	23.33	46.67	70.00	81.67	93.33	105.00	116.67	140.00	163.33	186.67	233.33
8	13.33	26.67	53.33	80.00	93.33	106.67	120.00	133.33	160.00	186.67	213.33	266.67
9	15.00	30.00	60.00	90.00	105.00	120.00	135.00	150.00	180.00	210.00	240.00	300.00
10	16.67	33.33	66.67	100.00	116.67	133.33	150.00	166.67	200.00	233.33	266.67	333.33
11	18.33	36.67	73.33	110.00	128.33	146.67	165.00	183.33	220.00	256.67	293.33	366.67
Days												
1	06	11	22	33	39	44	50	56	67	78	89	1.11
2	11	22	44	67	78	89	1.00	1.11	1.33	1.56	1.78	2.22
3	17	33	67	1.00	1.17	1.33	1.50	1.67	2.00	2.33	2.67	3.33
4	22	44	89	1.33	1.56	1.78	2.00	2.22	2.67	3.11	3.56	4.44
5	28	56	1.11	1.67	1.94	2.22	2.50	2.78	3.33	3.89	4.44	5.56
6	33	67	1.33	2.00	2.33	2.67	3.00	3.33	4.00	4.67	5.33	6.67
7	39	78	1.56	2.33	2.72	3.11	3.50	3.89	4.67	5.44	6.22	7.78
8	44	89	1.78	2.67	3.11	3.56	4.00	4.44	5.33	6.22	7.11	8.89
9	50	1.00	2.00	3.00	3.50	4.00	4.50	5.00	6.00	7.00	8.00	10.00
10	56	1.11	2.22	3.33	3.89	4.44	5.00	5.56	6.67	7.78	8.89	11.11
11	61	1.22	2.44	3.67	4.28	4.89	5.50	6.11	7.33	8.56	9.78	12.22
12	67	1.33	2.67	4.00	4.67	5.33	6.00	6.67	8.00	9.33	10.67	13.33
13	72	1.44	2.89	4.33	5.06	5.78	6.50	7.22	8.67	10.11	11.56	14.44
14	78	1.56	3.11	4.67	5.44	6.22	7.00	7.78	9.33	10.89	12.44	15.56
15	83	1.67	3.33	5.00	5.83	6.67	7.50	8.33	10.00	11.67	13.33	16.67
16	89	1.78	3.56	5.33	6.22	7.11	8.00	8.89	10.67	12.44	14.22	17.78
17	94	1.89	3.78	5.67	6.61	7.56	8.50	9.44	11.33	13.22	15.11	18.89
18	1.00	2.00	4.00	6.00	7.00	8.00	9.00	10.00	12.00	14.00	16.00	20.00
19	1.06	2.11	4.22	6.33	7.39	8.44	9.50	10.56	12.67	14.78	16.89	21.11
20	1.11	2.22	4.44	6.67	7.78	8.89	10.00	11.11	13.33	15.56	17.78	22.22
21	1.17	2.33	4.67	7.00	8.17	9.33	10.50	11.67	14.00	16.33	18.67	23.33
22	1.22	2.44	4.89	7.33	8.56	9.78	11.00	12.22	14.67	17.11	19.56	24.44
23	1.28	2.56	5.11	7.67	8.94	10.22	11.50	12.78	15.33	17.89	20.44	25.56
24	1.33	2.67	5.33	8.00	9.33	10.67	12.00	13.33	16.00	18.67	21.33	26.67
25	1.39	2.78	5.56	8.33	9.72	11.11	12.50	13.89	16.67	19.44	22.22	27.78
26	1.44	2.89	5.78	8.67	10.11	11.56	13.00	14.44	17.33	20.22	23.11	28.89
27	1.50	3.00	6.00	9.00	10.50	12.00	13.50	15.00	18.00	21.00	24.00	30.00
28	1.56	3.11	6.22	9.33	10.89	12.44	14.00	15.56	18.67	21.78	24.89	31.11
29	1.61	3.22	6.44	9.67	11.28	12.89	14.50	16.11	19.33	22.56	25.78	32.22
30	1.67	3.33	6.67	10.00	11.67	13.33	15.00	16.67	20.00	23.33	26.67	33.33
33	1.83	3.67	7.33	11.00	12.83	14.67	16.50	18.33	22.00	25.67	29.33	36.66
63	3.50	7.00	14.00	21.00	24.50	28.00	31.50	35.00	42.00	49.00	56.00	70.00
93	5.17	10.33	20.67	31.00	36.17	41.33	46.50	51.67	62.00	72.33	82.67	103.33

5000 Dollars.

Years.	½ per ct.	1 per ct.	2 per ct.	3 per ct.	3½ per ct.	4 per ct.	4½ per ct.	5 per ct.	6 per ct.	7 per ct.	8 per ct.	10 per ct.
1	25.00	50.00	100.00	150.00	175.00	200.00	225.00	250.00	300.00	350.00	400.00	500.00
2	50.00	100.00	200.00	300.00	350.00	400.00	450.00	500.00	600.00	700.00	800.00	1000.00
3	75.00	150.00	800.00	450.00	525.00	600.00	675.00	750.00	900.00	1050.00	1200.00	1500.00
4	100.00	200.00	400.00	600.00	700.00	800.00	900.00	1000.00	1200.00	1400.00	1600.00	2000.00
5	125.00	250.00	500.00	750.00	875.00	1000.00	1125.00	1250.00	1500.00	1750.00	2000.00	2500.00

Months.												
1	2.08	4.17	8.33	12.50	14.58	16.67	18.75	20.83	25.00	29.17	33.33	41.67
2	4.17	8.33	16.67	25.00	29.17	33.33	37.50	41.67	50.00	58.33	66.67	83.33
3	6.25	12.50	25.00	37.50	43.75	50.00	56.25	62.50	75.00	87.50	100.00	125.00
4	8.33	16.67	33.33	50.00	58.33	66.67	75.00	83.33	100.00	116.67	133.33	166.67
5	10.42	20.83	41.67	62.50	72.92	83.33	93.75	104.17	125.00	145.83	166.67	208.33
6	12.50	25.00	50.00	75.00	87.50	100.00	112.50	125.00	150.00	175.00	200.00	250.00
7	14.58	29.17	58.33	87.50	102.08	116.67	131.25	145.83	175.00	204.17	233.33	291.67
8	16.67	33.33	66.67	100.00	116.67	133.33	150.00	166.67	200.00	233.33	266.67	333.33
9	18.75	37.50	75.00	112.50	131.25	150.00	168.75	187.50	225.00	262.50	300.00	375.00
10	20.83	41.67	83.33	125.00	145.83	166.67	187.50	208.33	250.00	291.67	333.33	416.67
11	22.92	45.83	91.67	137.50	160.42	183.33	206.25	229.17	275.00	320.83	366.67	458.33

Days.												
1	07	14	28	42	49	56	63	69	83	97	1.11	1.39
2	14	28	56	83	97	1.11	1.25	1.39	1.67	1.94	2.22	2.78
3	21	42	83	1.25	1.46	1.67	1.88	2.08	2.50	2.92	3.33	4.17
4	28	56	1.11	1.67	1.94	2.22	2.50	2.78	3.33	3.89	4.44	5.56
5	35	69	1.39	2.08	2.43	2.78	3.13	3.47	4.17	4.86	5.56	6.94
6	42	83	1.67	2.50	2.92	3.33	3.75	4.17	5.00	5.83	6.67	8.33
7	49	97	1.94	2.92	3.40	3.89	4.38	4.86	5.83	6.81	7.78	9.72
8	56	1.11	2.22	3.33	3.89	4.44	5.00	5.56	6.67	7.78	8.89	11.11
9	63	1.25	2.50	3.75	4.37	5.00	5.63	6.25	7.50	8.75	10.00	12.50
10	69	1.39	2.78	4.17	4.86	5.56	6.25	6.94	8.33	9.72	11.11	13.89
11	76	1.53	3.06	4.58	5.35	6.11	6.88	7.64	9.17	10.69	12.22	15.28
12	83	1.67	3.33	5.00	5.83	6.67	7.50	8.33	10.00	11.67	13.33	16.67
13	90	1.81	3.61	5.42	6.32	7.22	8.13	9.03	10.83	12.64	14.44	18.06
14	97	1.94	3.89	5.83	6.81	7.78	8.75	9.72	11.67	13.61	15.56	19.44
15	1.04	2.08	4.17	6.25	7.29	8.33	9.38	10.42	12.50	14.58	16.67	20.83
16	1.11	2.22	4.44	6.67	7.78	8.89	10.00	11.11	13.33	15.56	17.78	22.22
17	1.18	2.36	4.72	7.08	8.26	9.44	10.63	11.81	14.17	16.53	18.89	23.61
18	1.25	2.50	5.00	7.50	8.75	10.00	11.25	12.50	15.00	17.50	20.00	25.00
19	1.32	2.64	5.28	7.92	9.24	10.56	11.88	13.19	15.83	18.47	21.11	26.39
20	1.39	2.78	5.56	8.33	9.72	11.11	12.50	13.89	16.67	19.44	22.22	27.78
21	1.46	2.92	5.83	8.75	10.21	11.67	13.13	14.58	17.50	20.42	23.33	29.17
22	1.53	3.06	6.11	9.17	10.69	12.22	13.75	15.28	18.33	21.39	24.44	30.56
23	1.60	3.19	6.39	9.58	11.18	12.78	14.38	15.97	19.17	22.36	25.56	31.94
24	1.67	3.33	6.67	10.00	11.67	13.33	15.00	16.67	20.00	23.33	26.67	33.33
25	1.74	3.47	6.94	10.42	12.15	13.89	15.63	17.36	20.83	24.31	27.78	34.72
26	1.81	3.61	7.22	10.83	12.64	14.44	16.25	18.06	21.67	25.28	28.89	36.11
27	1.88	3.75	7.50	11.25	13.13	15.00	16.88	18.75	22.50	26.25	30.00	37.50
28	1.94	3.89	7.78	11.67	13.61	15.56	17.50	19.44	23.33	27.22	31.11	38.89
29	2.01	4.03	8.06	12.08	14.10	16.11	18.13	20.14	24.17	28.19	32.22	40.28
30	2.08	4.17	8.33	12.50	14.58	16.67	18.75	20.83	25.00	29.17	33.33	41.67
33	2.29	4.58	9.17	13.75	16.04	18.33	20.63	22.92	27.50	32.08	36.67	45.83
63	4.38	8.75	17.50	26.25	30.63	35.00	39.38	43.75	52.50	61.25	70.00	87.50
93	6.46	12.92	25.83	38.75	45.21	51.67	58.13	64.58	77.50	90.42	103.33	129.17

6000 Dollars.

Years.	½ per ct.	1 per ct.	2 per ct.	3 per ct.	3½ per ct.	4 per ct.	4½ per ct.	5 per ct.	6 per ct.	7 per ct.	8 per ct.	10 per ct.
1	30.00	60.00	120.00	180.00	210.00	240.00	270.00	300.00	360.00	420.00	480.00	600.00
2	60.00	120.00	240.00	360.00	420.00	480.00	540.00	600.00	720.00	840.00	960.00	1200.00
3	90.00	180.00	360.00	540.00	630.00	720.00	810.00	900.00	1080.00	1260.00	1440.00	1800.00
4	120.00	240.00	480.00	720.00	840.00	960.00	1080.00	1200.00	1440.00	1680.00	1920.00	2400.00
5	150.00	300.00	600.00	900.00	1050.00	1200.00	1350.00	1500.00	1800.00	2100.00	2400.00	3000.00

Months.												
1	2.50	5.00	10.00	15.00	17.50	20.00	22.50	25.00	30.00	35.00	40.00	50.00
2	5.00	10.00	20.00	30.00	35.00	40.00	45.00	50.00	60.00	70.00	80.00	100.00
3	7.50	15.00	30.00	45.00	52.50	60.00	67.50	75.00	90.00	105.00	120.00	150.00
4	10.00	20.00	40.00	60.00	70.00	80.00	90.00	100.00	120.00	140.00	160.00	200.00
5	12.50	25.00	50.00	75.00	87.50	100.00	112.50	125.00	150.00	175.00	200.00	250.00
6	15.00	30.00	60.00	90.00	105.00	120.00	135.00	150.00	180.00	210.00	240.00	300.00
7	17.50	35.00	70.00	105.00	122.50	140.00	157.50	175.00	210.00	245.00	280.00	350.00
8	20.00	40.00	80.00	120.00	140.00	160.00	180.00	200.00	240.00	280.00	320.00	400.00
9	22.50	45.00	90.00	135.00	157.50	180.00	202.50	225.00	270.00	315.00	360.00	450.00
10	25.00	50.00	100.00	150.00	175.00	200.00	225.00	250.00	300.00	350.00	400.00	500.00
11	27.50	55.00	110.00	165.00	192.50	220.00	247.50	275.00	330.00	385.00	440.00	550.00

Days.												
1	08	17	33	50	58	67	75	83	1.00	1.17	1.33	1.67
2	17	33	67	1.00	1.17	1.33	1.50	1.67	2.00	2.33	2.67	3.33
3	25	50	1.00	1.50	1.75	2.00	2.25	2.50	3.00	3.50	4.00	5.00
4	33	67	1.33	2.00	2.33	2.67	3.00	3.33	4.00	4.67	5.33	6.67
5	42	83	1.67	2.50	2.92	3.33	3.75	4.17	5.00	5.83	6.67	8.33
6	50	1.00	2.00	3.00	3.50	4.00	4.50	5.00	6.00	7.00	8.00	10.00
7	58	1.17	2.33	3.50	4.08	4.67	5.25	5.83	7.00	8.17	9.33	11.67
8	67	1.33	2.67	4.00	4.67	5.33	6.00	6.67	8.00	9.33	10.67	13.33
9	75	1.50	3.00	4.50	5.25	6.00	6.75	7.50	9.00	10.50	12.00	15.00
10	83	1.67	3.33	5.00	5.83	6.67	7.50	8.33	10.00	11.67	13.33	16.67
11	92	1.83	3.67	5.50	6.42	7.33	8.25	9.17	11.00	12.83	14.67	18.33
12	1.00	2.00	4.00	6.00	7.00	8.00	9.00	10.00	12.00	14.00	16.00	20.00
13	1.08	2.17	4.33	6.50	7.58	8.67	9.75	10.83	13.00	15.17	17.33	21.67
14	1.17	2.33	4.67	7.00	8.17	9.33	10.50	11.67	14.00	16.33	18.67	23.33
15	1.25	2.50	5.00	7.50	8.75	10.00	11.25	12.50	15.00	17.50	20.00	25.00
16	1.33	2.67	5.33	8.00	9.33	10.67	12.00	13.33	16.00	18.67	21.33	26.67
17	1.42	2.83	5.67	8.50	9.92	11.33	12.75	14.17	17.00	19.83	22.67	28.33
18	1.50	3.00	6.00	9.00	10.50	12.00	13.50	15.00	18.00	21.00	24.00	30.00
19	1.58	3.17	6.33	9.50	11.08	12.67	14.25	15.83	19.00	22.17	25.33	31.67
20	1.67	3.33	6.67	10.00	11.67	13.33	15.00	16.67	20.00	23.33	26.67	33.33
21	1.75	3.50	7.00	10.50	12.25	14.00	15.75	17.50	21.00	24.50	28.00	35.00
22	1.83	3.67	7.33	11.00	12.83	14.67	16.50	18.33	22.00	25.67	29.33	36.67
23	1.92	3.83	7.67	11.50	13.42	15.33	17.25	19.17	23.00	26.83	30.67	38.33
24	2.00	4.00	8.00	12.00	14.00	16.00	18.00	20.00	24.00	28.00	32.00	40.00
25	2.08	4.17	8.33	12.50	14.58	16.67	18.75	20.83	25.00	29.17	33.33	41.67
26	2.17	4.33	8.67	13.00	15.17	17.33	19.50	21.67	26.00	30.33	34.67	43.33
27	2.25	4.50	9.00	13.50	15.75	18.00	20.25	22.50	27.00	31.50	36.00	45.00
28	2.33	4.67	9.33	14.00	16.33	18.67	21.00	23.33	28.00	32.67	37.33	46.67
29	2.42	4.83	9.67	14.50	16.92	19.33	21.75	24.17	29.00	33.83	38.67	48.33
30	2.50	5.00	10.00	15.00	17.50	20.00	22.50	25.00	30.00	35.00	40.00	50.00
33	2.75	5.50	11.00	16.50	19.25	22.00	24.75	27.50	33.00	38.50	44.00	55.00
63	5.25	10.50	21.00	31.50	36.75	42.00	47.25	52.50	63.00	73.50	84.00	105.00
93	7.75	15.50	31.00	46.50	54.25	62.00	69.75	77.50	93.00	108.50	124.00	155.00

7000 Dollars.

Years.	½ per ct.	1 per ct.	2 per ct.	3 per ct.	3½ per ct.	4 per ct.	4½ per ct.	5 per ct.	6 per ct.	7 per ct.	8 per ct.	10 per ct.
1	35.00	70.00	140.00	210.00	245.00	280.00	315.00	350.00	420.00	490.00	560.00	700.00
2	70.00	140.00	280.00	420.00	490.00	560.00	630.00	700.00	840.00	980.00	1120.00	1400.00
3	105.00	210.00	420.00	630.00	735.00	840.00	945.00	1050.00	1260.00	1470.00	1680.00	2100.00
4	140.00	280.00	560.00	840.00	980.00	1120.00	1260.00	1400.00	1680.00	1960.00	2240.00	2800.00
5	175.00	350.00	700.00	1050.00	1225.00	1400.00	1575.00	1750.00	2100.00	2450.00	2800.00	3500.00
Months.												
1	2.92	5.83	11.67	17.50	20.42	23.33	26.25	29.17	35.00	40.83	46.67	58.33
2	5.83	11.67	23.33	35.00	40.83	46.67	52.50	58.33	70.00	81.67	93.33	116.67
3	8.75	17.50	35.00	52.50	61.25	70.00	78.75	87.50	105.00	122.50	140.00	175.00
4	11.67	23.33	46.67	70.00	81.67	93.33	105.00	116.67	140.00	163.33	186.67	233.33
5	14.58	29.17	58.33	87.50	102.08	116.67	131.25	145.83	175.00	204.17	233.33	291.67
6	17.50	35.00	70.00	105.00	122.50	140.00	157.50	175.00	210.00	245.00	280.00	350.00
7	20.42	40.83	81.67	122.50	142.92	163.33	183.75	204.17	245.00	285.83	326.67	408.33
8	23.33	46.67	93.33	140.00	163.33	186.67	210.00	233.33	280.00	326.67	373.33	466.67
9	26.25	52.50	105.00	157.50	183.75	210.00	236.25	262.50	315.00	367.50	420.00	525.00
10	29.17	58.33	116.67	175.00	204.17	233.33	262.50	291.67	350.00	408.33	466.67	583.33
11	32.08	64.17	128.33	192.50	224.58	256.67	288.75	320.83	385.00	449.17	513.33	641.67
Days.												
1	10	19	39	58	68	78	88	97	1.17	1.36	1.56	1.94
2	19	39	78	1.17	1.36	1.56	1.75	1.94	2.33	2.72	3.11	3.89
3	29	58	1.17	1.75	2.04	2.33	2.63	2.92	3.50	4.08	4.67	5.83
4	39	78	1.56	2.33	2.72	3.11	3.50	3.89	4.67	5.44	6.22	7.78
5	49	97	1.94	2.92	3.40	3.89	4.38	4.86	5.83	6.81	7.78	9.72
6	58	1.17	2.33	3.50	4.08	4.67	5.25	5.83	7.00	8.17	9.33	11.67
7	68	1.36	2.72	4.08	4.76	5.44	6.13	6.81	8.17	9.53	10.89	13.61
8	78	1.56	3.11	4.67	5.44	6.22	7.00	7.78	9.33	10.89	12.44	15.56
9	88	1.75	3.50	5.25	6.13	7.00	7.88	8.75	10.50	12.25	14.00	17.50
10	97	1.94	3.89	5.83	6.81	7.78	8.75	9.72	11.67	13.61	15.56	19.44
11	1.07	2.14	4.28	6.42	7.49	8.56	9.63	10.69	12.83	14.97	17.11	21.39
12	1.17	2.33	4.67	7.00	8.17	9.33	10.50	11.67	14.00	16.33	18.67	23.33
13	1.26	2.53	5.06	7.58	8.85	10.11	11.38	12.64	15.17	17.69	20.22	25.28
14	1.36	2.72	5.44	8.17	9.53	10.89	12.25	13.61	16.33	19.06	21.78	27.22
15	1.46	2.92	5.83	8.75	10.21	11.67	13.13	14.58	17.50	20.42	23.33	29.17
16	1.56	3.11	6.22	9.33	10.89	12.44	14.00	15.50	18.67	21.78	24.89	31.11
17	1.65	3.31	6.61	9.92	11.57	13.22	14.88	16.53	19.83	23.14	26.44	33.06
18	1.75	3.50	7.00	10.50	12.25	14.00	15.75	17.50	21.00	24.50	28.00	35.00
19	1.85	3.69	7.39	11.08	12.93	14.78	16.63	18.47	22.17	25.86	29.56	36.94
20	1.94	3.89	7.78	11.67	13.61	15.56	17.50	19.44	23.33	27.22	31.11	38.89
21	2.04	4.08	8.17	12.25	14.29	16.33	18.38	20.42	24.50	28.58	32.67	40.83
22	2.14	4.28	8.56	12.83	14.97	17.11	19.25	21.39	25.67	29.94	34.22	42.78
23	2.24	4.47	8.94	13.42	15.65	17.89	20.13	22.36	26.83	31.31	35.78	44.72
24	2.33	4.67	9.33	14.00	16.33	18.67	21.00	23.33	28.00	32.67	37.33	46.67
25	2.43	4.86	9.72	14.58	17.01	19.44	21.88	24.31	29.17	34.03	38.89	48.61
26	2.53	5.06	10.11	15.17	17.69	20.22	22.75	25.28	30.33	35.39	40.44	50.56
27	2.63	5.25	10.50	15.75	18.38	21.00	23.63	26.25	31.50	36.75	42.00	52.50
28	2.72	5.44	10.89	16.33	19.06	21.78	24.50	27.22	32.67	38.11	43.56	54.44
29	2.82	5.64	11.28	16.92	19.74	22.56	25.38	28.19	33.83	39.47	45.11	56.39
30	2.92	5.83	11.67	17.50	20.42	23.33	26.25	29.17	35.00	40.83	46.67	58.33
33	3.21	6.42	12.83	19.25	22.46	25.67	28.88	32.08	38.50	44.92	51.33	64.17
63	6.12	12.25	24.50	36.75	42.88	49.00	55.13	61.25	73.50	85.75	98.00	122.50
93	9.04	18.08	36.17	54.25	63.29	72.33	81.38	90.42	108.50	126.58	144.67	180.83

8000 Dollars.

	½ per ct.	1 per ct.	2 per ct.	3 per ct.	3½ per ct.	4 per ct.	4½ per ct.	5 per ct.	6 per ct.	7 per ct.	8 per ct.	10 per ct.
Years.												
1	40.00	80.00	160.00	240.00	280.00	320.00	360.00	400.00	480.00	560.00	640.00	800.00
2	80.00	160.00	320.00	480.00	560.00	640.00	720.00	800.00	960.00	1120.00	1280.00	1600.00
3	120.00	240.00	480.00	720.00	840.00	1280.00	1440.00	1600.00	1920.00	2240.00	2560.00	3200.00
4	160.00	320.00	640.00	960.00	1120.00	1280.00	1440.00	1600.00	1920.00	2240.00	2560.00	3200.00
5	200.00	400.00	800.00	1200.00	1400.00	1600.00	1800.00	2000.00	2400.00	2800.00	3200.00	4000.00
Months.												
1	3.33	6.67	13.33	20.00	23.33	26.67	30.00	33.33	40.00	46.67	53.33	66.67
2	6.67	13.33	26.67	40.00	46.67	53.33	60.00	66.67	80.00	93.33	106.67	133.33
3	10.00	20.00	40.00	60.00	70.00	80.00	90.00	100.00	120.00	140.00	160.00	200.00
4	13.33	26.67	53.33	80.00	93.33	106.67	120.00	133.33	160.00	186.67	213.33	266.67
5	16.67	33.33	66.67	100.00	116.67	133.33	150.00	166.67	200.00	233.33	266.67	333.33
6	20.00	40.00	80.00	120.00	140.00	160.00	180.00	200.00	240.00	280.00	320.00	400.00
7	23.33	46.67	93.33	140.00	163.33	186.67	210.00	233.33	280.00	326.67	373.33	466.67
8	26.67	53.33	106.67	160.00	186.67	213.33	240.00	266.67	320.00	373.33	426.67	533.33
9	30.00	60.00	120.00	180.00	210.00	240.00	270.00	300.00	360.00	420.00	480.00	600.00
10	33.33	66.67	133.33	200.00	233.33	266.67	300.00	333.33	400.00	466.67	533.33	666.67
11	36.67	73.33	146.67	220.00	256.67	293.33	330.00	366.67	440.00	513.33	586.67	733.33
Days.												
1	11	22	44	67	78	89	1.00	1.11	1.33	1.56	1.78	2.22
2	22	44	89	1.33	1.56	1.78	2.00	2.22	2.67	3.11	3.56	4.44
3	33	67	1.33	2.00	2.33	2.67	3.00	3.33	4.00	4.67	5.33	6.67
4	44	89	1.78	2.67	3.11	3.56	4.00	4.44	5.33	6.22	7.11	8.89
5	56	1.11	2.22	3.33	3.89	4.44	5.00	5.56	6.67	7.78	8.89	11.11
6	67	1.33	2.67	4.00	4.67	5.33	6.00	6.67	8.00	9.33	10.67	13.33
7	78	1.56	3.11	4.67	5.44	6.22	7.00	7.78	9.33	10.89	12.44	15.56
8	89	1.78	3.56	5.33	6.22	7.11	8.00	8.89	10.67	12.44	14.22	17.78
9	1.00	2.00	4.00	6.00	7.00	8.00	9.00	10.00	12.00	14.00	16.00	20.00
10	1.11	2.22	4.44	6.67	7.78	8.89	10.00	11.11	13.33	15.56	17.78	22.22
11	1.22	2.44	4.89	7.33	8.56	9.78	11.00	12.22	14.67	17.11	19.56	24.44
12	1.33	2.67	5.33	8.00	9.33	10.67	12.00	13.33	16.00	18.67	21.33	26.67
13	1.44	2.89	5.78	8.67	10.11	11.56	13.00	14.44	17.33	20.22	23.11	28.89
14	1.56	3.11	6.22	9.33	10.89	12.44	14.00	15.56	18.67	21.78	24.89	31.11
15	1.67	3.33	6.67	10.00	11.67	13.33	15.00	16.67	20.00	23.33	26.67	33.33
16	1.78	3.56	7.11	10.67	12.44	14.22	16.00	17.78	21.33	24.89	28.44	35.56
17	1.89	3.78	7.56	11.33	13.22	15.11	17.00	18.89	22.67	26.44	30.22	37.78
18	2.00	4.00	8.00	12.00	14.00	16.00	18.00	20.00	24.00	28.00	32.00	40.00
19	2.11	4.22	8.44	12.67	14.78	16.89	19.00	21.11	25.33	29.56	33.78	42.22
20	2.22	4.44	8.89	13.33	15.56	17.78	20.00	22.22	26.67	31.11	35.56	44.44
21	2.33	4.67	9.33	14.00	16.33	18.67	21.00	23.33	28.00	32.67	37.33	46.67
22	2.44	4.89	9.78	14.67	17.11	19.56	22.00	24.44	29.33	34.22	39.11	48.89
23	2.56	5.11	10.32	15.33	17.89	20.44	23.00	25.56	30.67	35.78	40.89	51.11
24	2.67	5.33	10.67	16.00	18.67	21.33	24.00	26.67	32.00	37.33	42.67	53.33
25	2.78	5.56	11.11	16.67	19.44	22.22	25.00	27.78	33.33	38.89	44.44	55.56
26	2.89	5.78	11.57	17.33	20.22	23.11	26.00	28.89	34.67	40.44	46.22	57.78
27	3.00	6.00	12.00	18.00	21.00	24.00	27.00	30.00	36.00	42.00	48.00	60.00
28	3.11	6.22	12.44	18.67	21.78	24.89	28.00	31.11	37.33	43.56	49.78	62.22
29	3.22	6.44	12.89	19.33	22.56	25.78	29.00	32.22	38.67	45.11	51.56	64.44
30	3.33	6.67	13.33	20.00	23.33	26.67	30.00	33.33	40.00	46.67	53.33	66.67
33	3.67	7.33	14.67	22.00	25.67	29.33	33.00	36.67	44.00	51.33	58.67	73.33
63	7.00	14.00	28.00	42.00	49.00	56.00	63.00	70.00	84.00	98.00	112.00	140.00
93	10.33	20.67	41.33	62.00	72.33	82.67	93.00	103.33	124.00	144.67	165.33	206.67

125

9000 Dollars.

Years.	½ per ct.	1 per ct.	2 per ct.	3 per ct.	3½ per ct.	4 per ct.	4½ per ct.	5 per ct.	6 per ct.	7 per ct.	8 per ct.	10 per ct.
1	45.00	90.00	180.00	270.00	315.00	360.00	405.00	450.00	540.00	630.00	720.00	900.00
2	90.00	180.00	360.00	540.00	630.00	720.00	810.00	900.00	1080.00	1260.00	1440.00	1800.00
3	135.00	270.00	540.00	810.00	945.00	1080.00	1215.00	1350.00	1620.00	1890.00	2160.00	2700.00
4	180.00	360.00	720.00	1080.00	1260.00	1440.00	1620.00	1800.00	2160.00	2520.00	2880.00	3600.00
5	225.00	450.00	900.00	1350.00	1575.00	1800.00	2025.00	2250.00	2700.00	3150.00	3600.00	4500.00

Months.												
1	3.75	7.50	15.00	22.50	26.25	30.00	33.75	37.50	45.00	52.50	60.00	75.00
2	7.50	15.00	30.00	45.00	52.50	60.00	67.50	75.00	90.00	105.00	120.00	150.00
3	11.25	22.50	45.00	67.50	78.75	90.00	101.25	112.50	135.00	157.50	180.00	225.00
4	15.00	30.00	60.00	90.00	105.00	120.00	135.00	150.00	180.00	210.00	240.00	300.00
5	18.75	37.50	75.00	112.50	131.25	150.00	168.75	187.50	225.00	262.50	300.00	375.00
6	22.50	45.00	90.00	135.00	157.50	180.00	202.50	225.00	270.00	315.00	360.00	450.00
7	26.25	52.50	105.00	157.50	183.75	210.00	236.25	262.50	315.00	367.50	420.00	525.00
8	30.00	60.00	120.00	180.00	210.00	240.00	270.00	300.00	360.00	420.00	480.00	600.00
9	33.75	67.50	135.00	202.50	236.25	270.00	303.75	337.50	405.00	472.50	540.00	675.00
10	37.50	75.00	150.00	225.00	262.50	300.00	337.50	375.00	450.00	525.00	600.00	750.00
11	41.25	82.50	165.00	247.50	288.75	330.00	371.25	412.50	495.00	577.50	660.00	825.00

Days.												
1	13	25	50	75	88	1.00	1.13	1.25	1.50	1.75	2.00	2.50
2	25	50	1.00	1.50	1.75	2.00	2.25	2.50	3.00	3.50	4.00	5.00
3	38	75	1.50	2.25	2.63	3.00	3.38	3.75	4.50	5.25	6.00	7.50
4	50	1.00	2.00	3.00	3.50	4.00	4.50	5.00	6.00	7.00	8.00	10.00
5	63	1.25	2.50	3.75	4.38	5.00	5.63	6.25	7.50	8.75	10.00	12.50
6	75	1.50	3.00	4.50	5.25	6.00	6.75	7.50	9.00	10.50	12.00	15.00
7	88	1.75	3.50	5.25	6.13	7.00	7.88	8.75	10.50	12.25	14.00	17.50
8	1.00	2.00	4.00	6.00	7.00	8.00	9.00	10.00	12.00	14.00	16.00	20.00
9	1.13	2.25	4.50	6.75	7.88	9.00	10.13	11.25	13.50	15.75	18.00	22.50
10	1.25	2.50	5.00	7.50	8.75	10.00	11.25	12.50	15.00	17.50	20.00	25.00
11	1.38	2.75	5.50	8.25	9.63	11.00	12.38	13.75	16.50	19.25	22.00	27.50
12	1.50	3.00	6.00	9.00	10.50	12.00	13.50	15.00	18.00	21.00	24.00	30.00
13	1.63	3.25	6.50	9.75	11.38	13.00	14.63	16.25	19.50	22.75	26.00	32.50
14	1.75	3.50	7.00	10.50	12.25	14.00	15.75	17.50	21.00	24.50	28.00	35.00
15	1.88	3.75	7.50	11.25	13.13	15.00	16.88	18.75	22.50	26.25	30.00	37.50
16	2.00	4.00	8.00	12.00	14.00	16.00	18.00	20.00	24.00	28.00	32.00	40.00
17	2.13	4.25	8.50	12.75	14.88	17.00	19.13	21.25	25.50	29.75	34.00	42.50
18	2.25	4.50	9.00	13.50	15.75	18.00	20.25	22.50	27.00	31.50	36.00	45.00
19	2.38	4.75	9.50	14.25	16.63	19.00	21.38	23.75	28.50	33.25	38.00	47.50
20	2.50	5.00	10.00	15.00	17.50	20.00	22.50	25.00	30.00	35.00	40.00	50.00
21	2.63	5.25	10.50	15.75	18.38	21.00	23.63	26.25	31.50	36.75	42.00	52.50
22	2.75	5.50	11.00	16.50	19.25	22.00	24.75	27.50	33.00	38.50	44.00	55.00
23	2.88	5.75	11.50	17.25	20.13	23.00	25.88	28.75	34.50	40.25	46.00	57.50
24	3.00	6.00	12.00	18.00	21.00	24.00	27.00	30.00	36.00	42.00	48.00	60.00
25	3.13	6.25	12.50	18.75	21.88	25.00	28.13	31.25	37.50	43.75	50.00	62.50
26	3.25	6.50	13.00	19.50	22.75	26.00	29.25	32.50	39.00	45.50	52.00	65.00
27	3.38	6.75	13.50	20.25	23.63	27.00	30.38	33.75	40.50	47.25	54.00	67.50
28	3.50	7.00	14.00	21.00	24.50	28.00	31.50	35.00	42.00	49.00	56.00	70.00
29	3.63	7.25	14.50	21.75	25.38	29.00	32.63	36.25	43.50	50.75	58.00	72.50
30	3.75	7.50	15.00	22.50	26.25	30.00	33.75	37.50	45.00	52.50	60.00	75.00
33	4.13	8.25	16.50	24.75	28.88	33.00	37.13	41.25	49.50	57.75	66.00	82.50
63	7.88	15.75	31.50	47.25	55.13	63.00	70.88	78.75	94.50	110.25	126.00	157.50
93	11.63	23.25	46.50	69.75	81.38	93.00	104.63	116.25	139.50	162.75	186.00	232.50